The Disappearance of Butterflies

The Disappearance of Butterflies

Josef H. Reichholf

Translated by Gwen Clayton

polity

Originally published in German as *Schmetterlinge: Warum sie verschwinden and was das für uns bedeutet* by Joseh H. Reichholf © 2018 Carl Hanser Verlag GmbH & Co. KG, München

This English edition © 2021 by Polity Press
Reprinted 2020

The translation of this work was funded by Geisteswissenschaften International – Translation Funding for Work in the Humanities and Social Sciences from Germany, a joint initiative of the Fritz Thyssen Foundation, the German Federal Foreign Office, the collecting society VG WORT and the Börsenverein des Deutschen Buchhandels
(German Publishers & Booksellers Association).

Excerpt from *Vergängliche Spuren* by Miki Sakamoto reproduced with permission of Verlag Kessel. © Verlag Kessel, Remagen 2014. All rights reserved by and controlled through Verlag Kessel. www.forestrybooks.com

Excerpt from 'Blauer Schmetterling', from: Hermann Hesse, Sämtliche Werke in 20 Bänden. Herausgegeben von Volker Michels. *Band 10: Die Gedichte* © Suhrkamp Verlag Frankfurt am Main 2002. All rights reserved by and controlled through Suhrkamp Verlag Berlin.

Polity Press
65 Bridge Street
Cambridge CB2 1UR, UK

Polity Press
101 Station Landing
Suite 300
Medford, MA 02155, USA

ISBN-13: 978-1-5095-3979-6

A catalogue record for this book is available from the British Library.

Library of Congress Cataloging-in-Publication Data

Names: Reichholf, Josef, author.
Title: The disappearance of butterflies / Josef H. Reichholf ; translated by Gwen Clayton.
Other titles: Schmetterlinge. English
Description: Cambridge, UK ; Medford, MA : Polity Press, [2020] | Originally published in German as Schmetterlinge : Warum sie verschwinden and was das fur uns bedeutet by Joseh H. Reichholf 2018 Carl Hanser Verlag GmbH & Co. KG, Munchen. | Includes bibliographical references and index. | Description based on print version record and CIP data provided by publisher; resource not viewed.
Identifiers: LCCN 2020012872 (print) | LCCN 2020012873 (ebook) | ISBN 9781509539819 (epub) | ISBN 9781509539796 (hardback) | ISBN 9781509539796q(hardback) | ISBN 9781509539819q(epub)
Subjects: LCSH: Butterflies. | Butterflies--Ecology. | Butterflies--Conservation.
Classification: LCC QL543 (ebook) | LCC QL543 .R4513 2020 (print) | DDC 595.78/9--dc23
LC record available at https://lccn.loc.gov/2020012872 LC record available at https://lccn.loc.gov/2020012873

Typeset in 10.75 on 14 Adobe Janson by
Servis Filmsetting Ltd, Stockport, Cheshire
Printed and bound in Great Britain by TJ Books Limited

For further information on Polity, visit our website:
politybooks.com

Printed endpapers: nelsonarts/iStock

Contents

Foreword

Most people in the British Isles and North America – as well as on the European mainland – are by now well aware that many species of wildlife are declining and even disappearing. Butterflies and moths (Lepidoptera) are prominent among them. The author of this authoritative book, Dr Josef Reichholf, became concerned about the decline of many species in his native Germany during his doctoral and post-doctoral research at the Ludwig-Maximilian University of Munich on moths in both rural and urban areas around that city in the 1970s and 1980s. He became aware of the staggering speed of change and its extent, mainly due to the intensification of agriculture. The subject of his doctoral research was a group of aquatic moths belonging to the Crambid family, which he studied intensively in the field at a cluster of gravel pits near his boyhood home in Bavaria. Some time after completing his thesis, Dr Reichholf was saddened to find that some of these pits had been converted to swimming pools and others infilled and returned to agriculture, mostly for the cultivation of maize. At the same time, the riparian woodlands along the River Inn, which had been a favourite wildlife haunt of his – he started out as a bird-watcher in 1958 – were being uprooted to make way for the booming and profitable cultivation of maize. With the maize monoculture came the intensive use of pesticides to control the pests that infect it.

Dr Reichholf eventually made a long-term study of the Lepidoptera in urban and suburban Munich, mainly by employing light traps, and he came to the conclusion that, although the species diversity is less in urban areas, they are doing much better in the 'nature-friendly' cities than they are in the 'inhospitable' countryside, because the former are islands of warmth (the 'heat island' effect) and there is a much lower use of pesticides there. He devotes the second half of the book to a critical examination of the causes of the declines in so many butterfly and moth species: these, mostly familiar to British and North American nature conservationists, include the replacement of old, traditional farming by intensive agriculture, overuse of fertilizers and pesticides poisoning the soil, deforestation and reforestation with monocultures, habitat loss, over-tidiness, light pollution and the underlying effects of climate change.

Dr Reichholf holds strong views and is outspoken in his criticism of the role of politicians and government policies with regard to the environment and its wildlife. Some nature conservationists also come in for criticism for concentrating on legally protecting species from field naturalists and collectors and for not fully recognizing that their real enemy is industrialized agriculture. Finally, he considers the consequences of the disappearance of butterflies and moths and what can be done to arrest it.

This passionate, powerful and thought-provoking book could not be more timely. It deserves to be widely read by everyone concerned about the natural environment and I am very pleased to be able to recommend it.

John F. Burton, FZS, FRES

A Vice-President of Butterfly Conservation

Acknowledgements

I owe thanks to many people – too many to be able to list them all by name. The two main forces that led to the creation of this book came from my wife and my closest friends, on the one hand, and my publisher and agent, on the other. Working together with my literary agent Dr Martin Brinkmann and my editor Christian Koth of C. Hanser Verlag was both enjoyable and stimulating. The results make me feel optimistic, despite having experienced the decline of moths and butterflies over the last 50 years, that there is at least a glimmer of hope. If such a leading German publishing house can dedicate itself so closely to the subject of the disappearance of moths and butterflies, one must believe that there are still opportunities. I am very grateful to C. Hanser Verlag for conveying such optimism.

I thank my doctoral supervisor, the late Dr Wolfgang Engelhardt, for proposing the topic of aquatic moths and the extensive engagement in nature conservation that this led to, at a time when he himself was president of the Deutscher Naturschutzring. Through his suggestion, he effectively set the course of my professional life as a zoologist. Of central importance also was the Bavarian State Collection of Zoology (ZSM): a unique institution, where, as a member, one could feel 'butterflies' in one's stomach, while above, on the roof, real blue butterflies blithely flew around. It is beyond my abilities to try to put into words my gratitude for my time spent at the ZSM and my time as a teacher at both of the Munich universities, even though my essential attitude to life was formed professionally by these experiences. I am grateful to my wife, Miki Sakamoto-Reichholf, for the fact that I was able to combine them so well with my private life. She shares my enthusiasm for moths and butterflies.

It is a great pleasure to me to see my book translated into English, which opens my findings to an international audience. Some aspects, however, refer to the regional situation in Germany, especially in Bavaria and the adjacent regions of Austria. Others are of a much more widespread coverage. We are faced with the fact that the decline of butterflies, moths and other insects is a global phenomenon of our time. Many local and regional findings can be fitted together like a mosaic to create a picture that is already quite clear. It shows the continuing loss of biodiversity and natural richness.

I would like to thank Polity, especially Elise Heslinga, and the translator Gwen Clayton for their engagement. For me, it was a highly rewarding experience to work together with them in order to achieve a good translation. In my thanks I would also like to include John F. Burton for his highly valuable contributions to the English edition of my book. Last but not least, it is my hope that readers will be infected by the enthusiasm that I have felt throughout my years of research on moths and butterflies. This book is about their life and their future.

Introduction

In the last 50 years, our moth and butterfly populations have declined by more than 80 per cent. Perhaps only older people will recall a time when meadows were filled with colourful flowers and countless butterflies fluttered above. Nobody would have thought of wanting to count them then. Why would you! Butterflies belonged to summer, just like bees and wildflowers. Larks sang from early spring until mid-summer. They would sing from first light, suspended in the air over the fields. There were yellowhammers, partridges, hares. Frogs lived in the ditches and ponds. In the 1970s, treefrogs still called so loudly from a pool near my home at the edge of the fields that their chorus was audible through the veranda door during a telephone interview with the Bayerischer Rundfunk. The topic: proceedings at the Bavarian District Court regarding noise pollution caused by frogs.

I became familiar with butterflies when I was just a child. I saw dozens of large swallowtails with their distinct black lattice over pale-yellow wings. They flew to our vegetable garden to lay their eggs on carrot leaves. Their green caterpillars with red spots gave me particular pleasure when I discovered them weeks later. If I touched them near the front, they would shoot out the strangest orange-yellow fork from a wrinkle behind their head. They emitted a peculiar odour that I later learnt was a deterrent.

Blues of various species, which I could not tell apart at the time, flew over the meadows that stretched from our little house at the edge of the village to the woodland along the river. The shimmering blue butterflies were so abundant that, looking back, I could not even have estimated how many there were. One barely noticed the cabbage whites. They were part of the nature that surrounded us, like the chirruping of the field-crickets in May and June and the chirping of the grasshoppers in midsummer. I used to enjoy tickling the field-crickets out of their burrows with a stem of grass. Their bulky, brawny-looking heads amused me. There did not seem to be much going on in there – they were so easily tricked.

In the pollarded willows by the stream that snaked through the meadows behind our house, hoopoes would make their nests. With raised crests, they would stride around the grazed pastures, nodding their heads and poking around in the cowpats left behind by the cows. These birds were in the pastures during the day throughout the whole summer and far into the autumn. The air would teem with starlings. These black-feathered birds would follow the cows as if obsessed, sometimes even sitting on their backs. Every garden had at least one nestbox attached to a high pole for the starlings. When the cherries ripened, they feasted and took a significant share and made themselves heard in the process. Driving starlings away from the cherry trees was a great pleasure for older children, since they were allowed to climb right up into the tree crown, where the cherries dangled in front of their mouths. At our house, a colony of sparrows lived under the roof – a good dozen, maybe even more. They were always there, but our cat paid no attention to them. She went mouse-hunting and was very successful. A country idyll. Romanticized memories of childhood and early youth in a valley of the lower reaches of the River Inn, Lower Bavaria?

Perhaps nostalgia has affected our perception of the past. For this reason, one must be conscious of every attempt to reconstruct the 'former' as a basis for the 'present'. Memory supplies whatever we would like to have had, and it tends towards nostalgia and a yearning for what is gone. Nevertheless, I shall begin this book with descriptions of the natural beauty and abundance that I once experienced myself, in part to explain why the disappearance of the butterflies affects me so deeply. The first part of the book is intended to provide the basis from which we can make a judgement about the loss of the species. I have selected my examples so that the reader need not be a specialist in order to have observed and experienced similar things. These examples come from my own work and observations in Bavaria, but could equally have been taken from similar studies elsewhere in Europe or in the British Isles.

Together, these examples ought to show that the abundance of moths and butterflies, for reasons that are yet to be explained, has nevertheless shown a generally downward trend over at least the last 50 years (bearing in mind that it has always fluctuated substantially). The causes of this are discussed in the second part of the book. In order to do this, it is crucial to distinguish ordinary fluctuations from the general trend. This is critical, not only for understanding the natural cycles, but also for identifying the correct measures required to reverse the downward trend. It will not be achieved, for example, by simply reducing the application of poisons, as worthwhile as this might be. Whatever we commonly associate with 'green' and 'eco' holds its own problems with respect to the conservation of species. The second part of the book will therefore inevitably touch on environmental policy. The ecology movement lost its claim to scientific integrity, in my opinion, when it was converted into a 'nature religion' through crises that lent themselves to political manipulation. I am ready to be contradicted: I am used to this and it belongs to the principle of scientific discourse. Such discourse differentiates itself from the exchange of publicly entrenched opinions by accepting better findings. This makes natural science stronger, but also increasingly unpopular. It remains qualified and flexible, while people today seem to delight in dogmatically countering one principle with another. Scepticism does not disqualify you from being a natural scientist; instead, it is the praiseworthy habit of someone who does not submit to dogmas, even if they are currently supposed to be in fashion.

The same is true for the limitation of our freedom of expression under pressure from 'political correctness'. Whether we say 'plant protection products', as some demand, or 'poisons' does not change their effect, since that is what they are supposed to be: substances that kill what is supposed to be destroyed. Moreover, I have been unable to avoid writing in general terms of 'agriculture', 'maintenance measures' or 'nature conservation'. Farmers, if they so choose, can farm in an insect-friendly manner; a maintenance squad that cares for roadside verges can sometimes do this without mowing down all the grasses and flowers; and gardens can be designed in a very butterfly-friendly way. But the expressions 'agriculture', 'landscape and garden care' or even 'nature conservation', when referring to organizations and government works, are correct for the typical circumstances, since certain consequences emanate from them, and this book is generally concerned with these. For this reason, the impact of my statements will also depend on the spirit in which this book is read: I wrote it from a sense of responsibility that I feel we owe to future generations. Many people, a great many people, have been commenting on industrial agriculture for several decades, but they are still too few to achieve the political pressure that would be required to bring about a change for the better.

Part I

The Biodiversity of Lepidoptera

A Review of 50 Years of Butterfly and Moth Research

My records prevent me from creating rose-tinted memories of past conditions. I started keeping records on nature on 15 December 1958, and I therefore know, for example, that we did not have a 'white Christmas' in the Lower Bavarian Inn Valley 60 years ago, but it was, instead, two degrees above zero with light rain. On 2 January 1959, I noted that, outside on the River Inn, on the completely ice-free reservoir, I counted the following numbers of water birds: 800 mallards, 50 tufted ducks, 69 bean geese and 200 coots. Most of the numbers were rounded up, since I was not able to produce more accurate figures using my small binoculars from where I stood, half a kilometre away. I was not given a relatively powerful telescope until a few years later. As I pulled out and glanced through my old records in the course of preparing this book, I also came across a page with four butterflies that I had drawn myself. Astonished, I looked at it and read what I had written about the pictures 'drawn from my collection'. Next to a swallowtail and a pair of Adonis blues, *Polyommatus bellargus*, I had drawn a large and striking black and white butterfly, a great banded grayling, and labelled it with the scientific name that was customary at the time, *Satyrus circe*. This discovery surprised me, since the beautiful 'Circe' that flies in such an elegant manner has long vanished from my region. It is largely extinct in Southern

Bavaria, just like numerous other species of butterfly that I knew and observed in my youth.

Some of the species that were considered ordinary in those days do still exist, but they have, in the meantime, become rare or very rare. I also found a fitting example of this in my records. A note dated 12 September 1962 contained an observation that would be considered remarkable today. An almost palm-sized moth flew into the local train when we stopped at a station on our early morning journey to school and landed on the red shirt of my classmate. It was a red underwing, *Catocala nupta* (see Photo 1). When it is resting, the grey-brown, washed bark-coloured forewings of this large noctuid moth cover the bright crimson hindwings that are bordered with an angled black strip, just inside the outer margin. Not yet aware that moths – like all insects – cannot see the colour red, I wrote: 'The red underwing was thus attracted by the red colour of the shirt.' In fact, the red shirt would actually have seemed dark to the moth. To its vision in the so-called 'grey-scale', it may well have corresponded to a dark tree bark and the grey scales of its forewings. In the wild, the spot would have been suitable as a resting place during the day for this noctuid moth, which is active at twilight. One may therefore assume that red underwings were so common 50 years ago that one of them got lost in a train, presumably startled from its resting place in the station.

If this type of note from my schooldays was a mere anecdote, then these records would offer nothing further. Indeed, one retains what seems unusual, while the ordinary goes unnoticed. And yet, interesting points can be gleaned from unsystematic memos. I can find plenty of examples in my diaries. For example, the nine-spotted moth (or yellow belted burnet), *Syntomis phegea*, which has long since vanished from the area, sighted on 1 August 1960, or the caterpillar of the wood tiger, *Parasemia plantaginis*, recorded on 28 July 1960. The latter is only seen very rarely now. However, all of these and the many other records only show that there used to be Lepidoptera* species that no longer exist there. The true scope of the decline in butterflies and moths and other insects cannot be deduced from the disappearance of individual species. It is quite possible that other species that were not there earlier

* In other words, butterflies and moths. 'Lepidoptera' is an order of insects that includes both. [Tr.]

have appeared during this period. Nature is dynamic: changes can and will always occur. My initial claim that we have lost 80 per cent of the butterflies in the last 50 years refers to their overall frequency and requires much more concrete evidence.

I have already achieved that with birds: my counts of the water birds on the reservoirs of the lower River Inn, which I carried out every two to three days for six years, resulted in my first specialist ornithological publication in 1966. However, a quantitative survey of butterflies and moths was a very different challenge from counting birds that were resting on the banks or swimming on the water. My attempts gradually took form during my zoology studies at the University of Munich. A scientific approach was required for my doctoral thesis on aquatic moths, so I quickly familiarized myself with the five different species of moth that make up the Crambidae family and learnt how to reliably distinguish them by recognizing their flight patterns in the field.

However, given their number, moth and butterfly species require far greater knowledge if one wants to record all of them. The training is far more difficult and time-consuming than getting to know bird life. In southeast Bavaria alone, there are more than 1,100 species of butterfly and moth; for the whole of Bavaria, 3,243 have been reported (as of 2016). Many of these are very small and can only be identified with the help of specialist literature. For birds, there were already very good identification guides in the 1960s, which were not prohibitively expensive. Consequently, my initial engagement was with the bird world rather than with the butterflies. The reason was proximity, in the literal sense of the word: the reservoirs and riparian woods along the lower River Inn, which I could reach on foot or by bicycle, are a bird paradise. They are among the wetlands with the largest numbers of species in inland central Europe. When I started my zoology studies in Munich in 1965, I had already gained professional recognition as an ornithologist, thanks to my native surroundings, and I was familiar with various methods that are employed in field research.

Insects fly towards UV light

During my studies, I became familiar with a method that is more suitable than any other to establishing the abundance of moths. It consists of attracting species that are active at night using UV light. This is no longer done with large mercury-vapour lamps of 1,000 watts, which are used to light up white sheets that have been stretched behind them, as was common practice in the past and as I have attempted myself, but instead by means of an ingenious construction using UV neon tubes of only 15 watts. The moths and other insects are drawn in by this UV light. When they approach, they enter a funnel under the light tubes, leading to a large sack, in which the insects land. In order to offer them a place to hide until the following morning, empty egg cartons or similar are placed in the sack. This collection method does not harm the moths in any way. In the sack they quickly settle down, since the stimulus of the light has been removed. Together with the other insects, the moths are counted the following morning, and identi-fied species by species, to the extent that their species identification is possible. At that point, all the insects are released immediately. In this manner, readily analysable and statistically useable results are achieved, which can be compared, depending on the problem in question, with a similar assembly of the apparatus in another location. This method can even be used to establish frequency and species composition of moths in quite different habitats. This is exactly what I did, starting from 1969. Dr Hermann Petersen perfected the method. I am greatly indebted both to him, and also to Elsbeth Werner for allowing me to use her pesticide-free farm for research.

Unfortunately, this type of light attraction does not work with butterflies. In order to establish changes in frequency for them in a comparable manner, I started to count them in the 1970s, along specific, fixed routes that would not be changed over the years. For example, along forest tracks or dirt tracks across fields, or along embankments already specified as transects. With their signs indicating river kilome-tres, placed at intervals of exactly 200 metres, these riverside routes are perfectly suited to such transect counts. In the 1980s, I primarily used the findings that I had obtained in this way for my lectures on ecology and nature conservation at the Technical University of Munich and on ecological biogeography at the University of Munich. Over the years, it became clear that the light traps and the route counts (or 'transects') yielded fewer and fewer butterflies. As for the 'by-catch', as I referred to the other insects that flew into my lights, even the cockchafers disappeared, despite previously having been so numerous. On more than one occasion, their mass flight to the UV light caused the sack to become detached from the funnel and fall to the ground, since up to 1,000 cockchafers had crept in there in the gloaming. Since there were often hardly any moths in the cockchafer season at the beginning of May, such a mishap did not compromise the annual totals. But the rather sudden decrease in the cockchafers perplexed me. This was the first signal that my investigations were providing important data about the changes to nature. Yet in the 1980s, I still had no inkling of how sharply downhill things would go for the moths and the other insects, nor that my findings would result in an eco-nutritional basis for the decline of the birds in the meadows and the fields.

Urban Lepidoptera: more common than expected

In the early 1980s, I also started to research butterfly and moth frequency in the city. Munich, the place where I had been a scientist at the Bavarian State Collection of Zoology (ZSM), offered ideal conditions for this. There were enough suitable locations from the centre to the city margins to provide a kind of cross-section of the occurrence and frequency of nocturnal insects. Moreover, the large collections present in this research museum, together with the assistance of my specialist colleagues, stood at my disposal during my endeavours to identify precisely all the insects. The fact that I would need their help became evident as soon as I started work on my first findings. They were so much more species-rich than expected that I could never have coped on my own. Furthermore, the catches turned out to be extraordinarily rich in terms of quantity, too. The widely held idea that there would only be a pitiful fraction of the species diversity found in the countryside was not only called into question by my findings, but immediately exposed as mere prejudice.

Over the years and decades, the extensive research results that I shall report in this book thus came into being. They represent the results of half a century of quantitative entomology.

Over the past half-century, nature has changed to an extent and at a speed that are simply unprecedented in such a short period. The findings are staggering and the prospects that they imply are exceptionally grim. This is because we cannot expect the main agent of this loss of species diversity – agriculture – to undergo any substantial change. Anyone who delves into the 'agricultural problem' in any depth will find that it has less to do with the farmers themselves than with agricultural politics. The billions of subsidies they have received over the last 50 years have resulted in a highly competitive displacement of the small-scale farms by the large ones. Traditional farmers more or less disappeared, until only a tenth of their former numbers now remain, and yet the victor in this situation, international agrobusiness – in particular, the producers of crop protection products – managed to keep a low public profile, while the decline of insects and birds proceeded in shocking parallel to the death of small-scale farm-based agriculture.

The much-maligned city life has long since become better than life in the country, where the slurry stinks to high heaven and poison is used in unprecedented quantities, and where the birds have been silenced and the groundwater is no longer fit to drink. How can things go on like this? Is it not possible to curb the spirits we once called upon in good faith to lighten the work of farmers and improve their lives? Can we even imagine a 'butterfly effect' that might lead to a reversal in the state of industrialized agriculture? Although it may surprise you, my outlook at the end of this book is cautiously optimistic. And yet perhaps this hesitantly expressed optimism is nothing but a dream, for future generations won't appreciate what they haven't come to know or experience themselves: the biodiversity of nature and our moths and butterflies.

Death's head hawk-moth: a guest that can barely live with us anymore

Right now, species that could have had a lasting impact on our children, if only they had got to know them, are disappearing. I am thinking of the great wonder that seized me one evening in early October, when a death's head hawk-moth emerged from its pupa in a glass on my windowsill (see Photo 2). I had found it during the potato harvest. In the early 1960s, the potato fields in Lower Bavaria were predominantly harvested in the old-fashioned way. A horse pulled a plough with its share set at such an angle that, with the correct spacing, it created a deep furrow on one side, while a line of earth about two hand-widths higher than required was tipped over with the potato haulms. Some potatoes would thus escape harvest. We would look for these and then find others that were still encased in earth, which we would grub out by hand. In our Lower Bavarian dialect, we called this 'Kartoffelklauben' ('potato grubbing'). The caterpillars of the death's head hawk-moth, however, also now feed on potato plants in summer, since they are adapted to eating plants of the nightshade (Solanaceae) family. The potato plant, which originally came from South America, belongs to this poisonous plant family.

The distant origins of the death's head hawk-moth are not a disadvantage; they actually work in its favour. This is because there is nowhere in the wild where the females of these massive moths will find any nightshade plants in such abundance and so conveniently grown, with open ground around the bushes, as in a potato field. Accordingly, soon after its large-scale introduction into Europe in the early seventeenth century, the American potato plant became a preferred alternative for this large African hawk-moth. We can assume that, prior to this, it flew here from the edge of tropical Africa only seldom, due to the lack of suitable forage plants. The bittersweet nightshade, Solanum dulcamara, did not offer much nutrition and grows so sparsely that even today the caterpillar of the death's head hawk-moth is rarely found on it.

When these caterpillars are fully grown, they dig an elongated hollow slightly below the surface of the ground, in which they pupate. After several weeks of pupation, the mature moths emerge, and must try to fly south, across the Alps. They will not survive the winter to the north of them. My specimen was just such a fully grown caterpillar. I had housed him in a clean jam jar on a bed of garden soil and loosely covered him. From time to time I would spray the soil a little, so that the pupa would not dry out. With success: the newly emerged death's head hawk-moth, whose wings were not yet fully unfurled, not quite covering the yellow and black-ringed plump abdomen, appeared massive to me. More so, when I let it crawl onto my index finger so that I could place it on the curtains. This way of holding it would allow it to fully stretch out its wings, so that they could become fit for flying. My goal was, after all, to release it at dusk, so that it could return to Africa. The mere thought that this plan might be successful and that I might contribute to it excited me tremendously.

However, the pale-yellow design on the back of its thorax that was supposed to remind me of a skull made little impression on me. No matter which way I looked, it did not resemble a skull at all. Perhaps I simply lacked the imagination to conjure one up (see Photo 3). Even now, while I write this, I find it hard to imagine that men in earlier times could have come up with such an odd idea. But those who can see a liver in the leaves of a liverwort, just because it is made up of three 'lobes' (not remotely resembling a liver) and can conclude from this that it must be good for liver trouble, will surely manage to make out a mini-skull in the back markings of this hawk-moth.

I am no longer certain what was going through my head at the time, but something happened that awakened the biologist in me once and for all. Our cat, which had been lying on the sofa and pretending to be fast asleep, in the way that cats do, approached and stretched its nose up to the death's head hawk-moth on the curtains. The moment one of its whiskers touched the moth, the yellow and black-ringed abdomen flicked up between the wings and the moth let out a shrill squeak. The cat recoiled in shock, fell off the edge of the sofa, and took cover underneath it.

The combined effect of the wasp- or hornet-like markings and the squeaking sound, which extends to the ultrasound range, and which may thus be partially inaudible to the human ear, was an object lesson for me: no book, no account of such a procedure, could have explained in a more striking way what 'aposematic colouring' means and how it works. I was particularly affected, since I had found it so easy to coax the hawk-moth onto my fingertip and to place it on the curtains. Years later, whenever I stroked a bumblebee, coaxing it to produce a quiet humming, I would think back to this experience. This death's head hawk-moth, of which there would be dozens more in my life (until the potato harvest was mechanized and no more pupae were to be found or no specimen was able to survive the rough handling of the harvesting machine) – this moth that emerged under my care – touched me (see Photo 4).

The Fascinating Life of Aquatic Moths

Yet at that time I was even more interested in water birds. Shortly after starting my zoology studies, I drew together, and published, the results of my six-year study on their occurrence and frequency on the reservoirs of the lower River Inn. The work was formally accepted by the Zoological Institute as a thesis for the state examination, and I was therefore able to start looking for a topic for my doctoral thesis just two years after commencing my studies. It had to be a piece of field research, preferably an extension of my water-bird studies. However, the assistant professor whom I approached thought that I should save such an extensive research until after my PhD and that I would do better to prepare a topic that included problems of physiology. Of the three suggestions that he gave me, I liked the aquatic moths best. Their strange lives have much to do with physiology, in particular the physiology of breathing under water and the skin structure of the caterpillars, which enables life in water.

As my particular species, I selected the brown china-mark, *Nymphula nymphaeata*, now *Elophila nymphaeata*. It is thanks to this small, delicate moth that I gained my PhD (Dr. rer. nat. = Doctorate in Natural Sciences) at the Ludwig-Maximilian University in Munich in 1969. The title of 'doctor' is available for the most diverse areas of research, among them abundantly obscure topics whose usefulness is not immediately apparent to everyone. It is possible that my doctoral thesis, 'Research into the biology of the aquatic moth *Nymphula nymphaeata*', also falls into this category. In 1970 it was published in the *International Review of Hydrobiology* (volume 55, pp. 687–728). The fact that it dealt with the lifestyle of an aquatic moth with the charming name (when freely translated) 'little nymph of the water-lilies' is clear from the title. But understanding what this 'biology' actually consists of requires a more detailed description (see Photo 5).

This delicate moth and its close relatives still delight me so much that my heart beats noticeably faster when I see one of them or glean something new about their way of life. This still happens even though so much time has passed since I wrote my dissertation. In moments such as these, I think how lucky I was to choose this topic for my doctoral thesis. The little water-lily nymphs made me into a field biologist who would far rather carry out research outdoors in the natural world than indoors in a laboratory. They fired up my scientific enthusiasm for the beauty and the wonder of the living. I was never able to handle or look on the small moths and butterflies as 'objects', as mere research items, however scientifically productive. I was captivated by their vivacity.

How pleased I was when the small moths that I kept at home in an observation tank drank a droplet of lightly sweetened water from my fingertip. Afterwards they would always seem slightly confused and gaze around in bewilderment. The shallow bowl of water filled up with plants from the banks of small ponds – their natural habitat – that I had placed in the tank clearly did not appeal. I was keen to research their behaviour, but being held captive in this manner probably altered it quite significantly. The caterpillars of the aquatic moths could easily be kept in small aquariums. Their lives are focused entirely on the intake of food and on the regular moulting that is necessary for them to grow. During the feeding stage, they do not have any special requirements other than receiving the correct food in the form of leaves from floating-leaf plants. They do not even seem to be very particular about this, contrary to what was described in the specialist literature available at the time. I was therefore able to follow the development of the caterpillars, from the appearance of the eggs through to pupation and the emergence of the moths, directly from my workplace at the Zoological Institute of the University of Munich. But not the lives of the moths themselves.

Yet I was lucky: I came across the secrets of their lives in another, much better manner. Indeed, I first found 'my' aquatic moth in the Botanical Gardens in Munich, where the caterpillars devoured the floating leaves of small (and rare) water-lilies. They were rather unpopular with the gardeners for this reason. However, I did not free them of these pests through my research, as they surely hoped I would, since soon afterwards I found the water moths occurring naturally near my home village, in the Lower Bavarian Inn Valley. The brown china-mark moths, to give them their sober technical name, occurred in gravel pits that had been abandoned and even partially misused as rubbish dumps and in the oxbows of the riparian woodlands of the River Inn. I could reach them from home, either on foot or by bicycle. As I began my observations of the brown china-marks during those lovely evenings in early summer, I felt happy. As dusk started to fall, the males commenced their search flights. A first, ivory-coloured moth, easily recognizable in the evening light, danced out of the reeds along the bank and flew, like a wandering shadow, 20–30 centimetres above the water surface. It had barely appeared when dozens of others followed it out of the reeds. In the developing haze that very slowly thickened into mist, they danced their mysterious choreography. *Nymphula*, little nymph. What a fitting name, I thought, not yet suspecting that much greater treasures awaited me.

Evenings at the pond

As the sun went down, the croaking of the pool frogs rose to a final climax in polyphonic harmony, the noise feathering the surface of the gravel pit pond. Now it was the tree frogs' turn to start their part. But only one managed a short 'ep, ep, ep' and then went quiet. It was, after all, already too late in the year for the tree frogs. They give their concerts in April or at the start of May. Now rats investigated the rubbish heaps on the bank. Dark silhouettes, they scurried over rubble and household rubbish, looking for anything edible. For a few moments they distracted me from observing the aquatic moths. Looking through my low-light binoculars, I saw that there were brown rats of all sizes. There were some true giants, or so it seemed to me, that would have made deadly foes for cats. But there were also small ones that hunted around with their mothers and never strayed from their sides. As a trembling shadow flitted across the binoculars' field of vision, I noticed that there were also bats swirling around me. They were catching water insects over the pond. Twilight is the flight time for caddis flies and mayflies. As I pointed the focused beam of my torch this way and that they rose into the air in every direction. Late dusk turned to darkness. I had taken the torch with me in order to see for how long the aquatic moths flew. It seemed that they did not fly late into the night: in the last rays of daylight I recognized considerably fewer of them over the water surface.

The moths withdrew to the banks. From their short hovering flights, they landed on the stalks of plants and kept quiet. Perhaps it had become too cool for them, I thought, and felt for myself the moist, fresh chill of the early summer night. Based on these first impressions, I would have to measure the decrease in brightness and temperature. For the decrease in brightness, the light meter I still used to adjust the aperture of my camera in the 1960s ought to suffice. Measuring the air temperature would not be so easy, as I soon found out through attempts with a laboratory thermometer, since it showed very different temperatures depending on its proximity to the water surface, the reeds and the distance that I held it from my body. In field work half a century ago, we were a long way off today's precise temperature measurements. The counting of the flying moths was also rather problematic, to put it mildly. They swarmed so erratically over the water surface and along the edges of the reedbeds. In the dwindling evening light, they would gradually become more and more numerous, and then suddenly peter out. Counting attempts carried out rapidly one after another gave embarrassingly different figures. My conviction that I would be able to obtain an interesting doctoral thesis about these delightful moths using these methods gave way, over the next few evenings that I spent at the pond, to nascent anxiety as to whether I would be able to obtain findings that would be reliable and novel enough. Without doubt, it would still be necessary to keep the caterpillars in aquariums, together with the water plants they ate. The enclosures for the adult butterflies would have to be improved and redesigned so that they were closer to natural conditions. Obviously, conditions for making observations outside by the pond would not be ideal every evening.

It soon became apparent that, despite the convenience of having ponds with the aquatic moths almost by my front door, field research would have to contend with the weather and with unpredictable events. For example, a few weeks after the beginning of my investigations, one of the gravel pits was converted into a carp pond. Consequently, all the water plants there were immediately destroyed. At least back then, few of the fertilizers and poisons used in agriculture found their way into the gravel pits or the oxbow lakes in the riparian forest. I never suspected that a mere decade after the conclusion of my investigations, the pits would all be destroyed. They were filled in, levelled and planted with trees or turned back into arable land. Of all people, it was the nature conservationists who did this, declaring them to be 'wounds in the landscape' that had to be closed. New wounds of this kind would not be permitted, allowing the landscape to remain 'intact'. I still feel a pang whenever I go past the places where I carried out the field research for my doctorate. Some are covered by maize fields, while on one of them trees have grown and formed a small copse. They have all been filled in. Small losses such as these begin to erode one's sense of home. But back to the lifecycle of the brown china-mark. By continuing, I hope to show why these little beings fascinated me so much that they would affect me for the rest of my life as a zoologist, and why they became icons for me, for Lepidoptera in general.

The hidden lives of the little nymphs

Let us start with a successful search flight in early summer. It is the males who jiggle around in the evening twilight. They are searching for females, newly emerged ones that are still unmated. With a delightful little experiment, I was able to follow what was going on. Taking a female that had emerged in the aquarium at home, I put her into a mini cage of the type that beekeepers use to house queen bees for short periods. The attractant that the female produces escapes through the vented sides and is then dispersed by the air currents. Once the search flights of the males had started, I sent this virgin female off onto the open water of the pond on a polystyrene raft in the twilight. After just a few minutes, the erratic flights of the males became target-oriented trajectories. Dozens of males landed on the cage. With their abdomens stretched through the lattice of the cage, they attempted to reach the female. I had attached the little raft to a piece of string and could thus pull it towards me on the bank. The males followed as if drawn by a magnetic force They were not frightened away when I pulled out the stopper and released the female onto the floating leaves of the water knotweed or amphibious bistort, *Polygonum amphibium*. Barely was she freed, when one of the males captured her for copulation and would not let her go. The others did not have a chance. They left the pair and continued with their search flights. Here and there I noticed one or other of them landing and pairing with another female sitting on a floating leaf. The slightly more sombre-coloured females were not at all easy to spot in the evening half-light, as they sat on the floating leaves.

The next morning, the newly mated female begins her own search, this time for somewhere to lay her eggs. She flies ever closer over the surface of the water until she finds floating leaves without the notched edges that indicate that there are already caterpillars feeding there. She considers different species of water plants that spread their leaves on the water surface. After landing, the female carefully probes the leaf with her legs. Floating broad-leaved pondweed, *Potamogeton natans*, water knotweed (amphibious bistort), *Polygonum amphibium*, yellow floating heart (or fringed water-lily), *Nymphoides peltata*, and the young, fine leaves of water-lilies, *Nymphaea* sp. The floating leaves of the yellow floating heart (fringed water-lily), with its striking yellow flowers, are particularly suited for the caterpillars to grow and thrive. But this member of the bogbean family, which has evolved into a water plant, is only very rarely found. The variety of water plants mentioned here illustrates that the brown china-mark is not adapted to specific forage plants. Their caterpillars can actually be successfully fed on lettuce, admittedly with effects that I had not anticipated and that became very instructive.

When the female has found a suitable leaf, she pushes herself backwards towards its edge and curves the tip of her abdomen over it in such a manner as to reach the underside of the floating leaf. This is not at all easy for her, since she must overcome the resistance of surface tension. Close to the edge of the leaf, the female then sticks a cluster of 100–180 eggs on its underside. These are moistened by the water. The small caterpillars develop quickly. Precisely how quickly depends on how warm the water is during the early summer period. As they hatch, the caterpillars bite through the eggshell and gnaw their way into the tissue of the leaf. If the leaf is thick, as with the water-lily, then they will simply carry on eating; they 'mine', to be precise. With thinner leaves, they soon cut out a little piece and cover themselves with it. The finest insoluble silk threads hold the leaf segment, initially only 2–3 millimetres long, in place.

In this state, and in the next stage following the first moult, the caterpillar is moistened by the water in its tiny leaf tube. It breathes through its skin. The breathing holes, the spiracles and the trachea already exist, but remain closed in the first two stages of the caterpillar's life. Later on, air will be drawn through them into the body and the excreted carbon dioxide expelled. After the first moult, the caterpillar fashions for itself a proper leaf tube with base and lid. This also contains water and the caterpillar continues to breathe through its skin. However, this changes when it reaches the third larval stage. The caterpillar, having moulted the skin that had become too small, now shimmers as if made from silk. Water droplets run down its skin. It stretches its head above the surface of the water and is immediately enveloped in a gleaming silver layer of air. It fashions the new case of such a size that it can completely withdraw into it. It is filled with air. The body of the caterpillar will now remain hydrophobic until pupation. The spiracles are open. Now the exchange of respiratory gases takes place in the normal manner, but with a significant peculiarity: if the carbon dioxide level rises in the air bubble that surrounds the caterpillar, then a portion of it will automatically pass over into the water. This is because carbon dioxide 'readily' dissolves in water, as chemists will casually declare. The resulting lowered pressure is offset by the oxygen, which in turn penetrates the air bubble surrounding the caterpillar. The caterpillar thus breathes in part with something resembling a physical lung. This is not yet essential, since, from the third larval stage, the caterpillar of the brown china-mark eats the floating leaves from the top down. In doing so, it absorbs their waxes, which are responsible for ensuring that the top surface of the leaf does not become wet but instead floats on the water. Without this waxy surface, an ordinary rain shower would be enough to submerge the leaves.

How the caterpillar breathes under water

The addition of wax to the diet allows the caterpillar to move from the wettable to the unwettable (or hydrophilic to hydrophobic) next stage. Wax is secreted across the whole outer surface of the body of the caterpillar from tiny cone-shaped structures, the sides of which are deeply grooved and become filled with wax. That is why the caterpillars have such a silky sheen during their unwettable stage. Only the head and its connection to the body are not covered with such a wax layer. This is very important because if the head also repelled water, the surface tension would constantly push it away from the wet leaf and the caterpillar would barely be able to eat. The change from skin-breathing to air-breathing through the tubular system typical of insects, the spiracles, therefore characterizes the life of the caterpillar of this aquatic moth. It gets really exciting when the caterpillar is fully grown and ready for pupation. It does not crawl with its leaf case to land, although this might be the most practical solution from a human perspective, but instead labours downwards, along the stem of the water plant, struggling to overcome the buoyancy of the air-filled case. When it has reached a water depth of 10–30 centimetres, it bites a few small holes into the stem of the pondweed or water-lily plant, on the leaves of which it has been feeding, attaches the case by spinning a few fine threads, and transforms itself into a pupa. There, inside the air bubble, the pupa frees itself from the final caterpillar skin by light movements of its abdomen.

Then it rests until the internal transformation into a butterfly has been completed. Since this metamorphosis, which appears so quiet to the outside world, requires a great deal of energy, the pupa must breathe. This would lead to a shortage of air in the pupa case if the caterpillar had not tapped into the air tubes of the water plant. These tubes supply the roots of the plant with the oxygen they need, and the pupa obtains oxygen from them. Since carbon dioxide produced in the transformation process dissolves in the water that surrounds the leaf case, a lower pressure is created. This in turn causes air to be sucked down along the plant stem. In contrast to the caterpillar, which floats on the water surface in its air-filled vessel and can replenish its air directly, the pupa is dependent on the plant for its air supply. It is even possible that the leaves that remain green in the water carry out photosynthesis for longer, just to provide oxygen to the air bubble of the pupa. The highly complex problem of how an air-based animal can breathe under water is therefore solved in different ways: an adaptive achievement that one can only wonder at.

Up and away in an air balloon

The most impressive moment of all is when the moth emerges from the pupa. There is the moth, enclosed in the floating leaf case, 20–30 centimetres under water. It then pushes open the top of the leaf case. The escaping air bubble drags the moth up to the water surface, like a hot air balloon with its basket. At the surface the bubble bursts, and *Nymphula* emerges from the water. Immediately, a coating of long scales spreads out on the surface around the moth. Supported by the surface tension, it searches tentatively for the next leaf by probing with its legs. Once it finds a leaf, it crawls onto it and pumps up its wings until they are completely unfurled. The moth often emerges in the morning, but it can also hatch late in the afternoon and in the early evening. The newly hatched moths seek the cover of the plants on the bank as soon as they can fly.

There they attach themselves with their head pointing down, in what is for them a very characteristic pose. Seen from the water, as well as from the perspective of a bird hunting in the reeds for insects, this position conceals much of the shape of their body. One can only recognize that they are moths if one looks into the reeds at approximately the same level. But even then, the fine pattern of yellowish whorls and darker spots impedes visual detection. At the periphery of the reed bed, it remains damp enough on hot days for the small bodies not to dry out. Avoiding desiccation is particularly important for the males, as they must often spend several evenings and perhaps a whole week waiting to find a freshly hatched female that is ready to mate. It is less important for the females, since they die shortly after laying their eggs.

Only once have I been lucky enough to witness the emergence process in an aquarium. I already knew roughly how it would happen, since how else would the moth be able to reach the water surface from its underwater case? But as I saw it for myself, it was still as if a miracle had taken place.

When the adult emerges, it is midsummer. According to the type of weather experienced in spring, the second generation of brown china-marks hatches later in July or August, sometimes even at the beginning of September. What happens next? They must survive the winter. Overcoming this difficulty necessitates a special extra stage in the second lifecycle of the year. The caterpillars that are still wet from the water do not enter the third stage that would make them water-repellent in autumn, but instead crawl without a leaf case down the stem of the water plant to a depth of at least 30 centimetres below the water surface. There they bite a hole, provided the stem is thick enough, excavate a narrow, oblong cavity and draw themselves into this chamber, their shape bent crooked at one end like the handle of a walking-stick. In this manner they spend the winter months until April or the beginning of May, when the water plants begin to sprout new leaves. Any floating leaves still on the surface of the water die away late in autumn. The caterpillars hibernate inside the plant stems, even if ice forms on the water above them.

In the spring, the rising temperature of the pond water allows the plants to grow new shoots and leaves. This onset of growth evidently signals to the caterpillars that it is time to become active once more. They leave their cavities, crawl upwards and feed on the tender new leaves. This provides them with the wax necessary for their transformation into the water-repelling condition. By May, the only caterpillars will be those in air-filled leaf cases, feeding hungrily on new floating leaves until they are fully grown and ready for pupation. The moths that hatch from those pupae make up the first generation. Their descendants will continue to develop, without the need for a hibernation period. With these two reproductive cycles, the year of the 'little nymph', *Nymphula*, is complete.

The advantages of living in water

My dissertation also dealt with the details of the skin structure and its changes during the period when the caterpillar stops breathing through its skin and starts to breathe oxygen using the system of trachea that is typical of insects. This required imaging with an electron microscope, which the University of Munich was able to arrange. But what was truly exciting for me was – and remains – the lives of these moths with their adaptations to the water plants that sustain them and their life in the water. Why did they come to inhabit this environment? What advantages does it offer them?

The discovery that would be of greatest importance in addressing this issue did not even occur to me at first: nearly all my attempts to breed caterpillars were successful and produced moths. Indeed, nothing could happen to them in my small aquarium, except perhaps damage through my own carelessness. All the pupae that I collected outdoors (together with their underwater leaf cases) in order to watch the emergence process emerged successfully. Without giving it any thought, I assumed that all the caterpillars in their various stages that I had collected for my research would continue to develop without any problems, pupate and produce moths. The penny only dropped, as the saying goes, years later, when I had already become involved with a quite different type of species, the small ermine moth. There is a separate chapter devoted to them. Through them, the advantage of life in the water became suddenly apparent: I had had no losses, because the caterpillars and pupae of my aquatic moths had not been attacked by parasites. For practically all the butterflies and moths that live on dry land, parasites are among the main factors that determine their abundance and their development from one stage to the next. With around 96–98 per cent of 694 caterpillars from several breeding groups, the hatching success of my aquatic moths was phenomenally high. I only recorded higher losses for the eggs. I did not discover who or what caused the losses under outdoor conditions, but I considered the egg-eating water mites and the rotting sludge build-up in the heavily silted pools to be the likely causes. With 100 or more eggs per clutch and per female moth, such losses prevent the caterpillars from consuming all the available floating leaves too soon, which can easily happen where brown china-marks exist in large numbers.

This was why the gardeners in the Botanical Gardens in Munich placed their hopes in my research into the aquatic moths when I first encountered the little nymphs. Over the following years and decades, I definitively established that the female brown china-mark will leave the pool from which she emerged if the floating leaves of the water plants have been overconsumed. She will examine the edges of the floating leaves quite thoroughly before laying her eggs, and for good reason. If there is extensive feeding damage, she will leave and search for other waters with better conditions. A tendency to disperse would already be expected, since such small bodies of water are normally only temporary. Under natural conditions, they arise through inundation of the floodplains. New ponds will last a couple of years or a few decades, depending on how large or small they are when they form, and gradually disappear again through sedimentation and plant succession. Species that colonize an environment that is by its nature unstable must seek alternatives in good time.

The dispersal behaviour of the aquatic moths is therefore very particular. As insects, they probably belong to the group of pioneering species that is familiar to us through many land-based plants and that quickly colonizes newly created environments. On the other hand, maybe we are dealing with specialists that need a specific, longer-lasting 'life zone': that of floating leaves at the edges of large bodies of still water. Closer to the centre of the body of water there are plants that grow entirely under water, described by specialists as 'submerged'. The moths seek out shore plants that stand in the water but protrude above it, away from the centre of the pool. These are plants that are 'emerged' (in the ecological sense). In order to understand my aquatic moth and to be able to place it properly among its relatives, I would need to deal with the environment of small waterbodies and shores in far greater detail. Was it a pioneer species or was it specially adapted to the specific environment of bodies of water?

A place to live or an 'ecological niche'

The gravel pits combined all the conditions of the larger waterbodies in a small space. Accordingly, in most of them, I was also able to find the various other species of aquatic moth that exist in central Europe. They form an almost exemplary 'ecological series', feeding variously on the shore (or 'emergent') plants, through the floating leaves and all the way down to the submerged water plants. This sequence of adaptation is visible in the moths themselves. The caterpillars of the beautiful china-mark, *Nymphula stagnata* (*nitidulata*)* feed on bur reed, *Sparganium* sp. and other species of plant that project out of the water near the bank. Moving out into the water, to the zone where plants with floating leaves grow, is 'my' *Nymphula nymphaeata*. The series continues under water with the ringed china-mark, *Paraponyx stratiotata*, and the most unusual of all, *Acentropus niveus*. In side pools, provided they are covered with duckweed, you will also find the small china-mark, *Cataclysta lemnata*. All these species live next to one another, in the strictest sense of the word, in adjacent ecological niches. All have special adaptations that, in the extreme case of the water veneer, *Acentropus niveus* (*Acentria ephemerella*), with its two forms of female, have even led to a permanent life in the water. More about this shortly. First, the bigger picture must be understood. It shows the diversification of moths belonging to the small moth family, the Crambidae, known to be extraordinarily adaptable, from the bank right out into the water. The further they have advanced, the more abundant they are.

The abundance of a species is, in a general sense, an indication of its biological success. The caterpillars of *Nymphula stagnata* live to a greater or lesser degree on the banks, above the water surface. They are the rarest species in our series. *Cataclysta lemnata*, whose caterpillars use the small leaves of the duckweed plant to construct their cocoons and for nutrition, usually becomes considerably more common as one moves towards the centre of the body of water, but its occurrence is limited to small waterbodies that are carpeted in duckweed plants, which are also known as 'water lentil', or *Lemna*. The occurrence of my little nymph, *Nymphula nymphaeata*, is much more widespread and frequent. In order to build their leaf cases, its caterpillars cut out a pair of oval leaf sections, up to 3 centimetres in length. This can be readily seen from land.

In contrast, the ringed china-mark is much harder to detect. Its caterpillars spend their whole lives under water. They do not pass through an air-breathing stage. They do not have to, since they develop thread-like appendages on their bodies, through which the respiratory gas exchange takes place just like in the gills of fish. They are aptly

---

* The changes to the genus names and their current form are explained on page 112. They are hopefully now permanent.

called spiracular gills. Such gills are a very unusual adaptation for moths, although they are normal for the larvae of a very species-rich group of true water insects, the caddis flies. This raises a question that is eminently important for an understanding of the evolution of the Lepidoptera, namely whether spiracular gills are an invention by a genus of aquatic moths, or an ancient legacy linking moths with caddis flies. In other words: are moths and butterflies descended from formerly water-dwelling insects, or early forms of insect that were already established on dry land? There is much to indicate a closer relationship with caddis flies. In any event, they were all 'born from the water', just like my little *Nymphula*.

With a delicate rocking motion, the caterpillars of the ringed chinamark pump water through the loose cocoon, in which they sit under water, eating aquatic plants – in Germany, this is principally water-milfoil, *Myriophyllum* sp. To do this, they have adapted to survive in warm, oxygen-poor water. In the tropics, these aquatic moths, whose caterpillars develop spiracular gills, have a species-rich network of relationships. However, the *non plus ultra* of our aquatic moths spends its life as a caterpillar in the depths, among the massive stands of underwater plants that can grow as far as the water surface to flower. It is the tiny water veneer, *Acentropus niveus* (*Acentria ephemerella*), which the lepidopterists of the nineteenth century did not even recognize as a moth, taking it instead for an unusual species of caddis fly.

The caterpillars of the water veneer moth are comparatively normal. Their bodies are wettable and they breathe through their skin. Small as they are, they do not require a more efficient gas-exchange mechanism. Skin respiration is quite adequate for them, even in the cool and oxygen-rich shallow lakes in which they are principally found. They pupate under water. Yet what emerges from some of the pupae seems barely credible: females whose wings have been shortened to form pointed paddles. With these, they 'fly' around under water. Not fast, but fast enough to avoid simply being pushed to the surface. Their hind legs are fringed with a thick row of bristles and they can use these to steer. Stub wings and rudder legs enable these females to achieve goal-oriented movement under water. The reason for this is evident soon after they have emerged from the pupa. They struggle upwards to the water surface, remaining below it, and thrust only the tip of their abdomen out of the water. Glands on that tip will emit a scent that attracts the male with its normal wings.

The males will swoop around just above the water surface as if entranced until they encounter the abdomen tip of a female that is ready to mate. In the course of the coupling, they are almost pulled into the water by the larger female, but their wings prevent them from being dragged down into the deep. When the sperm has been delivered, the male releases the 'clamp' with which it gripped the tip of the female's abdomen. She, in turn, crawls and paddles down and looks around, 'flying' and 'paddling' until she finds a water plant that is suitable for egg deposition. I have found the caterpillars of this aquatic moth on curled pondweed, *Potamogeton crispus*, water-milfoil, *Myriophyllum* sp., and, above all, on Canadian waterweed, *Elodea canadensis*, which in the 1960s and 1970s was still relatively common in lakes and larger lagoons among the reservoirs along the Lower River Inn.

Having said this, those females with wings adapted to paddling are rare in southern Germany, in contrast to, for example, Denmark, southern Scandinavia and Britain. In central Europe, the females usually develop with normal wings. They are significantly larger than the males – and this is essential. This is because it is only the females, travelling on the wing, even if they are borne and blown along by the air currents rather than by actually flying, that are in a position to find bodies of water in which there are perfectly suitable stocks of underwater plants. The pools found near rivers and reservoirs have existed for too short a time to be considered permanent. However, the much more constant shallow lakes around the Baltic Sea have also only been there since the last ice age, that is, for around 10,000 years. If this aquatic moth had produced only females with rudimentary wings that were unable to fly, this species would surely not have survived in the long term. The males need their flight capability in any event, in order to search for females.

At this point I would like to mention a genetic peculiarity. With butterflies and moths, the female sex is genetically marked XY and the male XX; that is, the exact opposite from us. For this reason, it is much easier for females to develop two different forms than it is for males. This characteristic manifests itself with particular clarity in forms of 'mimicry', that is, through the imitation of poisonous or foul-tasting examples by non-poisonous imitators that are not protected by unpleasant flavours. In short: in moth and butterfly circles, it is worthwhile for the females to be more highly diversified. We can see this clearly with the female of the brimstone butterfly, which resembles the poisonous cabbage whites. The females, with the precious cargo of eggs in their bodies, have good reason to keep themselves concealed. Generally speaking, we therefore see considerably more male butterflies and moths in nature than females.

But let us return to the species composition of the aquatic moths. With their ecological placement on the banks of waterbodies and their lives on and in small bodies of water, they not only provide a prime example of how the species are distributed across their specific habitats (each in their own 'niche'), but also illustrate why the evolution of all the related adaptations in physique and lifestyle are so rewarding. The water edges constitute an environment rich in plants and luxuriant with plant stock, less affected by the vicissitudes of the weather. Plentiful nutrition is always attractive; plants that are not protected by special toxic substances all the more so. Vegetation right at the water's edge and, above all, under water, is particularly attractive, not only due to its utility but also because it is situated in a place that the main enemy, the parasitic insects, have difficulties in reaching. The life and survival of any type of moth or butterfly almost always depends on the success of its caterpillars. If the caterpillars do not find sufficient food plants, then that species will not do well. If there is enough food, but the caterpillars feeding on it are heavily parasitized, then the species will not become (more) abundant. One such case is found among those butterflies with caterpillars that eat nettles. Since nettle plants are in plentiful supply, these particular butterflies should also be extraordinarily numerous. They are indeed numerous, but not exceptionally so, and their populations fluctuate from year to year. More of this in a separate chapter that will provide an insight into the nature of fluctuations. With regard to the aquatic moths, there is yet another question: what do they teach us about this general trend, the disappearance of butterflies?

The destruction of the biotopes of the little nymphs

The detailed portrayal of their lifestyle set out above could give the impression that aquatic moths only have such a major role in my book because my research into butterflies and moths started with them. However, they serve as an illustration for the precarious situation in which so many butterflies and moths have ended up. As I shall explain in more detail below, the water-filled gravel pits in which they lived when I carried out my doctoral research no longer exist. They have been buried, filled in. Those that were in the open fields have been reincorporated into agricultural land. They belong to the large group of small structures of which farming land has been 'cleansed'. In this manner, the aquatic moths and many other butterflies and moths lost the biotopes in which they had flourished for many decades. Dozens of moths and butterflies – and, if one includes the smaller species, surely more than a hundred, as well as hundreds of other insects – made these gravel pits into islands of species diversity at the time when the homogenization of the fields was just beginning. In the opinion of certain influential nature conservationists at the time and of those authorities with jurisdiction over the excavation sites, gravel extraction had to be regulated and concentrated in special cultivation areas that should, of course, be as large as possible. When gravel extraction was complete, they could be repurposed as swimming ponds or, if of largely dry construction, could be recultivated as arable land or woods. It was this same trend that moved away from small- and medium-scale enterprises towards large businesses, and which could simultaneously be seen in agriculture.

With my studies of aquatic moths, I was only able to record a narrow spectrum of the insects that lived in gravel pits and small waterbodies. I observed their colony loss through the disappearance of the ponds in which dragonflies and frogs, and in dryer areas also lizards and beetles, had been comfortable. One by one, hedgerows and field copses also disappeared. The riparian woods by the River Inn were almost totally uprooted because of booming maize cultivation: the land was now worthless for growing trees, since firewood was no longer in demand, but had become highly profitable for farmers if they planted maize. The ban on woodland clearance came too late. Large parts of the riverine woods had already been destroyed and were further cleared over the years because the authorities did nothing to prevent it from happening. Perhaps it was only thanks to a series of wet summers that the wetter riparian woods along the River Inn remained untouched. Financial incentives for clearing woodland were eventually stopped.

It is with very mixed feelings that I think back to the 1970s when the fight for these alluvial forests played itself out and nature conservationists like me could only watch helplessly as piece after piece was cleared and converted into maize fields. The media and the general public were not very interested at the time. Such species-rich areas were referred to in the official parlance as 'wasteland' or 'barren land', descriptions that say everything about the value placed by large-scale agriculture upon animals and plants that have no direct commercial use. Sometimes I doubt whether our attitude to what is truly noteworthy has changed at all, even though one seldom hears talk of wasteland anymore. Words are interchangeable; it is far harder to change their underlying meaning, and there are setbacks time and again. Decades of serious attempts to replace the terms 'raptor' and 'predator' in Germany with 'bird of prey' and 'carnivore' (on the basis that 'prey' only exists in the eyes of the hunter) were fruitless. For some reason, young journalists in Germany have started using the term 'raptor' again quite openly, even though it disappeared from ornithological books long ago. 'Pest' is once again used for insects, regardless of whether or not they cause damage intentionally or even whether they cause any damage at all.

My aquatic moths caused no damage in the wild, in the small bodies of water or on lake shores. Unfortunately, the same could not be said of waterlily ponds. Water-lily species with thin leaves, in particular, were devoured by caterpillars, resulting in nothing more than an eyesore. In a garden pond with small, pink-flowering water-lilies most people want flawless floating leaves and, if possible, flowers that have not been spoilt by aphids. But anything that interferes with their look counts as damage, and for many an owner of such garden ponds it would be better if the little moths did not exist at all. Tropical aquatic moths also cause severe harvest losses in rice fields. The opportunity offered to me by the FAO (Food and Agriculture Organization) to carry out my investigations there did not attract me. The moths and butterflies that lived around us were interesting enough. Little was known in the 1970s about moths, apart from the forest pests. Nevertheless, it is the nocturnal moths, not their diurnal relatives the butterflies, that make up by far the largest proportion of the Lepidoptera species. Light itself would shed light on the darkness.

The Benefits of Being Attracted to Light

They used to be called 'sun children',* flying creatures of the air that flutter here and there, apparently aimless yet self-sufficient, letting themselves be borne on the wind and seeking cover at the sight of the first clouds, since they cannot tolerate the shade. This romantic view is most likely still true for some species of butterfly. The world of Lepidoptera, however, in all its richness and diversity of species and behaviour, only really unfolds when twilight falls. Since we are diurnal, many of us would be surprised to know that there are more than ten times as many species of moth than there are of butterfly. It is only when moths ghost around the streetlights on a sultry summer night that we suddenly become aware of them.

Many caterpillars also prefer to feed in darkness. This protects them from the birds that hunt them by day. It is harder for the adult insects to escape bats, since they hunt at night. It is true that their ultrasound does not work as well in the thick undergrowth of leaves, twigs and branches. Moths that fly in the open air at night are much more likely to be caught. Certain species of moth have developed the quite extraordinary capacity to detect the ultrasonic pulses emitted by bats without having 'heard' them. Then, if they are struck by them, they let

* In German, *Sonnenkinder* – see also p. 147.

themselves fall into the grass or the bushes, as quick as lightning. This is quite hard for us to grasp. To offer a plausible comparison, if we had this capacity, we would be able to detect radar traps when driving and brake in time to avoid them, without being conscious of having done so. Night-flying moths are, above all, concerned with finding a flight path through the darkness without crashing into obstacles if, somewhere, perhaps several hundred metres away, a newly emerged female is exuding her own kind of sexual lure. The travelling males remain unscathed, which is surprising considering that they are practically flying blind with only the scent-sensors in their antennae to steer with. How they manage this never ceases to amaze me, since it is light that steers them off their course and causes them to bump into obstacles. 'More light, worse sight' is hard for us to comprehend, since we depend so much on vision.

Night-flying moths avoid the day and pass the hours of light resting under cover. They start to come alive towards the end of dusk, when the weakening natural light forces us humans to go to sleep or to use artificial light. Yet it is that very light that attracts almost all moths, as if by some magical force. Certain wavelengths are particularly effective: above all, the 'weak' ultraviolet light (UV-A) that is invisible to us. Moths can see it, just like many other insects and the majority of birds, since they have an optical pigment that is sensitive to the wavelength of UV light. Birds that hunt at night, however, such as owls, do not use UV light, but the remaining daylight from the visible spectrum (as we would). Some owls, such as the barn owl, can hear the squeaking of a mouse that extends far into the ultrasound range, and can use this to locate that mouse and accurately swoop down to catch it, but only if there is sufficient residual light for them to do so: they avoid flying in complete darkness. The moths and night-flying butterflies evidently also need such residual light. But based on the current level of knowledge, we can barely even begin to speculate how they process this and how they use it to manage their often quite rapid flights, since, in terms of structure and function, their eyes are not significantly different from those of the butterflies and day-flying moths.

Like moths to a flame

The best nights for 'light-trapping' are the darkest ones with no moon, thick cloud cover and a light (warm) rain. Clear moonlit nights, on the other hand, are not very good. One may assume from this that moths orientate themselves by the moon and thus manage to achieve a relatively straight flightpath, since they can maintain a fixed bearing on the moon's distant light. But even if this were enough to maintain a particular direction, for example on a migration north or south, as is undertaken by some migrant moths and butterflies (for example, the death's head hawk-moth and other hawk-moths), it is of little help in avoiding collisions with leaves and branches in the woods or the garden. The fact that they are attracted by artificial light is only partially explained with reference to the moon, if at all. If one watches the moths as they fly towards the light, one will rarely see a straight, goal-oriented approach. Even the alleged spiral, with which they are supposed to approach light sources, is the exception rather than the rule, if it exists at all. And when is the moon ever so readily visible?

The fact is that light does attract a great many moths, but we do not yet understand why. On the other hand, we do know what type of light is particularly effective. Accordingly, the bulbs used in street lighting should be such that they emit light frequencies that either do not attract moths at all, or only barely – that is, above all 'yellow' and not 'blue' and in any event light that is as free of UV rays as possible. The much-maligned light pollution that occurs at night could at least be constituted more thoughtfully. We should not use any artificial light with the same spectral composition as sunlight, even where lighting is reasonable and necessary, such as in the illumination of roads for the safety of pedestrians and drivers. The same is true for floodlights used to light up historic buildings and other attractions and display windows. A more favourable spectral composition of artificial light would also benefit people, since we know that blue light is disturbing (as it should be, if it is used by police, fire engines and ambulances). Blue light disturbs moths too, and distracts them from their flights in search of animals of the same species for mating or the correct plants for egg-laying. We see this when moths swarm around lights on humid summer nights.

The red blindness of butterflies

The UV component is most probably of cardinal importance. It is a form of high energy radiation that interacts with visual pigments as well as with protective pigments such as the melanin in our skin. We know that tanning is the simplest, most visible biological reaction of our bodies to light. The skin cells are stimulated by the UV to produce protective melanin. When the skin becomes brown, it regulates the extent to which light can affect the skin. In all probability, the UV light vision of insects and other creatures is based on a reaction of living cells to a process caused by light. Incidentally, the fact that we cannot see UV light is more likely to be due to a loss of an earlier, more primal ability.

This limitation, and our severely reduced night vision, are the price we probably paid for developing trichromatic colour vision, through the capacity to recognize red and distinguish it from green. Lepidoptera cannot see red and red light does not attract them. Not only do they see very differently from us at night, but their whole view of the world, transmitted through the sensory organs, differs very significantly from ours. This is one of the more profound reasons why we find it more difficult to become attuned to the basic needs of butterflies and other insects than to those of birds. With their colour vision, birds are actually closer to us than most mammals, from whose world we have evolved. Dogs and cats are unimpressed by the beautiful red colour that we love. Together with green, it creates a mixed colour of inferior quality, of which people who are red-green colour blind could give the best description, inadequate though it would be, since red is not available as an experience to them. The alluring effect of UV on butterflies and moths is comparable to the attraction of red for most people.

We have this characteristic to thank for the best findings on the occurrence and abundance of many nocturnal insects, since they fly towards UV-rich light. What is so beautiful and appealing in individual butterflies lies in the eye of the beholder, that is, in our eyes. They do not see each other like this at all, and birds also recognize them in other ways. On their night-time hunts, bats estimate where they are using sonar imaging, which is different again. Yet over the thousands and millions of years of their existence, butterflies and moths must have learnt to deal continually with the visual ability of birds and the echoes produced by the ultrasound of bats. The challenges presented by humans, on the other hand, are still relatively new. A century of nights illuminated by electric light is not enough. Or so one might think.

But perhaps we should not take such a pessimistic view. There is actually a rich diversity of species of nocturnal moths living in those big, bright cities. Light pollution in general cannot therefore be the main factor determining their occurrence and abundance. More about this when we discuss the findings of the 'light catches' in the cities. But if the spectral composition of the artificial light sources were to be gradually adjusted to become more insect-friendly, this would certainly significantly improve the living conditions for the Lepidoptera in the cities. Then there would not be just 'small moths', but also hawk-moths the size of small birds and emperor moths that could be mistaken for small bats, flying ghost-like around the gardens and buildings of the city. Bats have long since fared much better in cities than in the countryside, where almost nothing of interest for them moves across the fields at night. The differences in numbers are enormous; the trends alarming. Thanks to the attraction to light, we know how things are for moths and many other insects of the fields. They will therefore occupy a central place in the second part of this book. Here I would like to continue with another, highly peculiar attraction, one that can be observed out in the forest without any technical assistance at all.

The Strange Behaviour of the Purple Emperor

The purple emperor, with its tropical blue sheen, is surely among the most impressive of the butterflies that live in our part of the world. There are two species in central Europe, the large purple emperor, *Apatura iris*, and the lesser purple emperor, *Apatura ilia*, which is only somewhat smaller than the large one. The males of both species shimmer a magnificent indigo, just as if the upper side of their wings were underlaid with silver. With the females, the indigo is slightly less deep or virtually missing. There is also a special red variant of the lesser purple emperor, whose wings shimmer a dusky pink. It is rare for us, however, to come close enough to these masterpieces of our butterfly world to be able to admire them in all their living beauty, since they fly quickly and usually at quite a height. In Britain, the male purple emperor, *A. iris*, is referred to as 'His Majesty'. For the many butterfly lovers there, sighting him represents the highlight of the butterfly year. They drive from near and far to the relatively few places in southern England where this treasure can be found with certainty.

In central Europe, both species of purple emperor can still be found quite regularly in early summer, between the end of June and the middle of August. The lesser purple emperor usually appears slightly earlier than its larger relative, usually from around mid-June. In particularly warm summers, such as 2018, one might even come across them at the beginning of June. Warmth always affects them. The main flying period for the larger purple emperor is the transition from June to July. The smaller species prefers the moister, warmer riparian woods, while the larger species favours logged forests – in southeast Bavaria, even forests dominated by spruce. If you did not know any better, you would be surprised to find the purple emperor in this type of managed woodland, which looks more like a plantation than a natural forest. But they are there, together with their smaller relatives, so that I need to inspect them closely in order to see which species they are when I set out to count the purple emperors along the forestry tracks. Often, I can already recognize the males of the larger purple emperor at quite a distance based on their distribution along the track. They sit right in the middle, at quite regular intervals of 20–30 metres – or every half kilometre, depending on how common or scarce they are in that particular year. From time to time, they fly up and perform an elongated oval over the forestry track, first in the direction of their neighbour on one side, then towards the other, which can sometimes lead to short aerial battles. Then the regular distribution pattern of the males is reproduced for a time. In other words, they claim a section of track as a territory that they defend against other members of their own species as well as against the lesser purple emperor, although less forcefully. If they are not close enough, they seem to be just as bad at differentiating these from their own species as we humans are. There is a marked, tooth-shaped point in the white stripe on the underside of the wing of the larger purple emperor, which disappears when the wings are closed, and which one can only make out when one is quite close. Similarly, the lesser purple emperor has another small ochre-coloured 'eye' on the lower outer edge of its forewing, and not just on the hindwing like the purple emperor.

I have always struggled to get close enough to this jewel of a butterfly to attempt a photograph: whenever I looked through the viewfinder of the camera and had the impression that I was close enough, had chosen the right section of the view and found the proper angle to capture the heavenly sheen, it would disappear with a barely perceptible wingbeat. Annoyed, and also ashamed of my slowness, I would later ascertain that on many such supposed butterfly pictures there would be nothing but the bare surface of the forestry track. Anyone who wants to take good photographs of such flighty beauties will need a lot of patience. Even then, it will barely suffice if you are not familiar with the behaviour of the butterfly and have no idea how it is influenced by the prevailing air temperatures or by the interplay of light and shadow in the forest. Early morning is the best time for photographs, when the sun is quite high and the butterflies are still numb from the cool and damp of the night. That said, at that time of day our purple emperors are usually still sitting in the unreachable heights of the treetops. Sometimes, one is simply lucky, and a photo comes out well. At other times, there might be special circumstances, such as those I once came across on a perfectly ordinary summer day in early July.

Butterflies on drugs

That morning, my wife and I were taking the dog for a walk in the forest. It was still cool enough for him under the trees. As usual, he walked a couple of metres ahead of us and sniffed his way carefully along the edges of the forestry track. When a few large butterflies flew up into the air, disturbed by the tip of his snout, he stopped short and sniffed the spot, but it appeared to hold nothing of great interest for his nose. He turned instead to trees where other dogs were accustomed to leaving their scent marks. We had not yet moved far from the parking area by the edge of the forest, and were therefore in the zone in which dogs that are taken to the woods mark most thoroughly. I would not have paid any further attention if the butterflies had not immediately flown back to the same place. Only then did it occur to me that our dog must have poked them directly with his nose. That was certainly highly unusual, since the group consisted of different butterfly species, namely two red admirals, a large purple emperor and two lesser purple emperors – that is, butterflies that were usually especially elusive. At 25°C in the shade, it was definitely warm enough for them, but what struck me was the peculiar flight pattern of these butterflies. Barely had they been roused than they settled down again on the same spot with their wings closed, that is, brought together vertically, so that when seen from above they looked like nothing more than five straight lines. Photographing lines is not particularly interesting, even when the lines are actually butterflies. But recording the camouflage effect of the folded wings piqued my scientific interest. When else does one get such an opportunity, and, what is more, with three different species?

I also photographed a side view of the butterflies, so that they would be more recognizable. The fact that I did not need to slither up to them commando-style across the dusty forestry track, given that they were undisturbed by my approach, astonished me even more. One does not let such rare opportunities slip through one's fingers. I photographed them from increasing proximity and hoped that they would do me the additional favour of opening their wings. But this they did not do, or at least, if they did, it was so instantaneous and so incomplete that I had no chance to capture the moment. Then, suddenly, one of them, a lesser purple emperor, put me in a most awkward situation. I must have got too close to it with my camera lens, wanting to see exactly how and where it was poking its unfurled proboscis. It flitted onto the camera and from there onto the back of my hand. There it sat and unfolded its wings so that they were perfectly flat, and their brightness shone right into my eyes.

For seconds I simply stared, enjoying this unparalleled view of the magnificent butterfly on my hand, and then, as slowly as I could, I took the camera with my left hand and gave it to my wife, so that she could capture this unbelievable moment. My movements did not upset the butterfly. Tangibly clinging onto my skin with its legs as I cautiously turned my hand, it allowed itself to be coaxed into a position that was more convenient for me and for the photo. Now it sat with outstretched wings right on the palm of my hand. With its lemon-yellow proboscis, it touched my fingers, a delicate but insistent contact that I distinctly felt. In the photograph, the antennae, proboscis and the position of the legs demonstrate that this butterfly is alive and has not been doped for an exhibition picture (see Photo 6). After several shots, which were difficult enough, since the camera had to capture the correct angle to make the emperor's sheen visible, it fluttered away.

Then I noticed that a few paces away a large purple emperor was simply sitting in the middle of the forestry track. He, too, allowed himself to be photographed from just a few centimetres away, although with his wings folded up. The red admiral, which was also still there, did exactly what the lesser purple emperor had done: when I was almost close enough to touch it, it, too, flew onto my hand. Such behaviour is certainly not normal for butterflies. What caused it? My hands were neither sweaty nor dirty. Why did the purple emperor with the folded wings just sit there? I had never seen such a thing. The fact that some butterflies readily suck perspiration is interesting, but nothing special. My wife and I had experienced this repeatedly when bathing by unspoilt, flower-dense banks, where blues flew around in great numbers. One must simply remain still for long enough, for example in the semi-shade, and use no sunscreen, which would disturb the butterflies' sense of smell. Occasionally we would make a game of it, allowing the blues to land on particular parts of the body. A fingertip dipped in perspiration, for example, would be suitable, or a big toe. Having made oneself attractive, one must approach the blue very slowly and with great care, so as not to frighten it away; it is a delightful game of patience on a hot and sultry summer's afternoon.

It is also well known that some butterflies are attracted by fermenting tree sap. Butterfly collectors would make use of this in earlier times, with their own highly confidential recipes. These would almost always contain a little alcohol mixed with fruit, or overripe cheese, such as a Limburger or other potent cheese that would have smelled quite strong even to our noses. They would smear this onto promising sections of tree trunks or other exposed places, marking them just like dogs. Experienced collectors would confirm this, since they know that certain butterflies can actually be attracted to dog excrement or urine in clayey or sandy places. Lesser purple emperors in particular will swarm around dog excrement, but will usually fly away immediately if you approach. However, animal excrement, perspiration and alcohol could be excluded as causes of the strange behaviour of my five butterflies on the forestry track, as could overripe cheese. The cause was actually a carcass, as I then saw, and quite a flat one, about the size of a hand. I had seen butterflies feasting on dead animals before; most recently, a white admiral on a dead roe deer fawn that lay in the woods and was in a state of advanced decay. My attempt to photograph the butterfly that was feeding on it was ultimately successful, but required me to hold my breath, since the deer smelt so foul. Because the white admiral kept flying away, I had to advance several times before I managed to get close enough. Despite all this, it was not a striking picture, and I was forced (grudgingly) to admit it was little more than a useful record.

Still, the carcass I was now seeing, which was a squashed toad, did not offer an obvious explanation for the curious behaviour of the butterflies. The common toad was quite desiccated and anything but succulent. And yet, the five butterflies dabbed repeatedly with their proboscises, as if patting the coarse skin down (see Photo 7). Butterflies can emit small droplets of liquid from the proboscis in order to dissolve mineral substances from the ground and absorb them. They would surely have had more success on the dusty track, especially at its edge, than on the back of the dead toad. And yet there was doubtless something particularly attractive about the toad. I picked it up and placed it at the side of the track, in the hope that the butterflies would not be run over there. They barely reacted. The red admiral even allowed himself to be relocated sitting on my hand, his wings semi-folded.

I very carefully grasped the purple emperor that was resting nearby by the wings, which it held steadily folded, and placed it on the safe roadside. It was apparent that there was nothing wrong with it, since it unrolled its proboscis, which had been rolled into a disc, briefly felt the air with it, and rolled it up again. Its legs formed a steady base, holding its body and its wings upright. Everything seemed to be in order, except that it was neither startled nor prompted to fly away. Its wings were undamaged. The butterfly was surely not too old or too weak to survive.

There seemed to be nothing wrong with the other butterflies either. And yet they behaved equally strangely. I was perplexed, and remained so until, after several days had passed and I had repeatedly looked at the pictures that I had taken, a possible explanation for their outlandish behaviour occurred to me. Poison! The butterflies had absorbed a toxin from the carcass of the common toad. They were numb, stoned, high – call it what you will. The toxin had essentially switched off their sensory impressions. They were no longer aware of what was going on around them and what I was doing with them. Almost addicted, they attempted to continue feeding on the cadaver. The large purple emperor had perhaps already absorbed too much and could not go on. But it still extended its proboscis. The butterflies restlessly felt around on my hand with the tips of their proboscises, and allowed themselves be moved by me, since there was no toad toxin on my hand. The toxin is produced in the two large glands that the common toad has on its head, just above the eyes. It prevents snakes, such as the grass snake, from

eating toads that they have already seized if they are not particularly
hungry. Toad poison was once an essential ingredient of recipes that
'witches' brewed in the late middle ages and in the early modern period
in order to 'fly': psychedelic drugs. Toad toxins in the right dosage will
have this effect.

Psychedelics in the insect kingdom

Various other insects also contain substances that have a psychedelic
effect. Take, for example, the famous (or rather, infamous) Spanish fly,
Lytta vesicatoria. It is not actually a fly but a beetle, related to the oil
beetle. 'Spanish fly' is still sold as an aphrodisiac. The Medici family
are thought to have become rich through the distribution of a special
tincture known as *aqua tofana*, which, in the right dosage, strengthens
virility. The active ingredient, cantharidin, has since become well known.
It is more effective than potassium cyanide, corrodes the mucous membranes and destroys the kidneys. The toxicity level of this slow-moving
beetle prevents it from being eaten by birds. Even greater protection is
afforded by the toxin carried by the larvae of the *Diamphidia nigroornata*,
or Bushman arrow poison beetles, in southern Africa. The bushmen of
the Kalahari manufactured a lethal arrow poison from them in former
times. Many Lepidoptera are 'poisonous' or 'bitter-tasting' because
their caterpillars ate plants containing toxic substances or precursors
to toxins that were converted into toxins inside their bodies through
the process of digestion. Particularly poisonous moths, such as the
diurnal Zygaenidae (burnet and forester moths), which are related to
the tiger moths, fly so slowly that one can catch them quite easily. It is
assumed that they do this on purpose. But cause and effect are probably
the other way around. They have to fly slowly because, if they became
too active, the toxins in their bodies would start to affect them. Toxins
will always constitute a dangerous cargo, but they can also represent an
opportunity for success, as can be seen from the large whites and their
population, which continues to be comparatively high. I shall deal with
this in a separate chapter.

As for what the butterflies do with the toad toxin that makes them
so torpid, I do not know. When they feed on alcohol from rotting
fruit, which clearly also intoxicates them, we are faced with the same

question. Yet, since alcohol can be broken down and converted into energy, one can at least imagine (or convince oneself) that this justifies the risk of being caught by a bird because they are tipsy and unfit to fly. Perhaps we are too reluctant to accept the idea that animals also look for drugs simply because they are effective, and not because they have a particular biological purpose. But what sort of drugs are contained in a pile of dog turd? When butterflies swarm around this, it is even harder to understand.

An experience I had with a lesser purple emperor at the beginning of June 2018 is pertinent here. We had just finished evaluating a light trap. The morning sun was streaming over the roof of the small, remote farmstead and we were drinking a cup of coffee to round off our very productive work: the trap had attracted 168 moths of 59 different species and a dozen other insect species too. A purple emperor flew towards us out of the wood. When the light caught its wings at the right angle, they seemed to light up, like those of the much larger, more azure, morpho butterflies from the tropical rainforests in Central and South America. It was therefore a male, and, as we saw when it briefly rested on the house wall, of the lesser purple emperor species (see Photo 8). It swung in elegant arcs past the gold and brown immortelle flowers as if carrying out test flights, looped the loop around the petunias, whose flowers in their abundance of bright colours hung from the window sills and from the balcony of the house, putting on a show worthy of a picture postcard, and flew behind the house only to reappear again immediately. It flitted around, here and there, this way and that, for nearly five minutes, and then floated past the balcony and disappeared from sight, until we discovered the object of its desire. It was not any particular flowers or a sunny spot with just the right temperature. It was the cat litter tray. It was not easy to locate this, since there was barely any wind that morning. But the butterfly found it, with its erratic and almost chaotic style of flying, and was finally able to enjoy the excrement of the cat, which slunk around the table by our feet and watched the moths with us, as we checked the catch from the previous night.

What do these anecdotes have to do with the disappearance of the butterflies? Purple emperors are, if I may be permitted the expression, glimmers of hope. They still exist, and indeed, as I would like to emphasize, in greater, not smaller numbers than in the forests of my earlier research activities. Taking into account the fact that I have visited the riparian woods and commercial woodland of southeast Bavaria significantly more often since my retirement, because June and July were always term-time at the universities of Munich, and I cannot keep away from the purple emperors on a favourable day, I would tend to say that their population was 'unchanged'. This is a helpful finding, since one must clarify which butterflies and moths in which habitat types have become scarcer in order to discover the causes. For this reason, we will now turn our attention to the Lepidoptera of the nettle patch, which have a few things in common, not least that nettles are the principal host plants for their caterpillars. Our considerations will be focused on the caterpillars and their requirements rather than on the butterflies themselves.

The Nettle-feeding Lepidoptera: An Instructive Community

Peacock butterflies, small tortoiseshells, red admirals and large cabbage whites are our best-known butterflies. Practically everyone has heard of the peacock or the white. But why are they so well known? Surely not just because of their beauty, although the peacock is an impressive butterfly (see Photo 9). One would not rank the large white among the especially beautiful, in any event. The reason for their familiarity is as simple as it is significant. We are ourselves responsible for their abundance. To put it plainly, they are synanthropic. Large whites are considered pests when their caterpillars eat up every type of cabbage in the fields or in the garden. The cultivation of cabbage has offered this butterfly, which thrives on plants from the species-rich cabbage family, a lavish food supply. The large white used to occupy the position that is today occupied by the European corn borer, *Ostrinia nubilalis*, which would demolish two and half million hectares of maize in Germany summer after summer were it not held in check by highly effective insecticides. Peacock butterflies, small tortoiseshells and red admirals, on the other hand, can benefit from quite different plants that have unintentionally been encouraged by humans, particularly through the use of agricultural fertilizers: nettles. Together with other less well-known butterflies and moths, the large whites make up the ecological group that we might term 'nettle Lepidoptera' or 'nettle-feeding Lepidoptera'.

Nettles: indicators of overfertilization

If we take a closer look at the lives of these butterflies, it is easy to understand one of the very major changes that have taken place in nature in the last few decades. The keyword here is *overfertilization*. The principal plant involved here is the nettle, or, to be more precise, the two species of nettle: the common nettle, *Urtica dioica*, and the annual, small or dwarf nettle, *Urtica urens*. The Latin name of the former highlights an important distinction: *dioica* means 'two houses'. This means that there are distinct male and female plants of this species of nettle. *Urens* in the latter means 'burning': it stings more painfully than its larger relative. With the annual nettle, the male and female flowers develop together on the same plant.

The fact that this botanical peculiarity is not without significance will become clear in due course, but there is another important difference between the two species of nettle. The common nettle develops a widely branching root network underground. Such a quantity of nutrients is stored here over the winter that it can swiftly grow new plants in the spring. The common nettle is perennial, while the small nettle is annual; in other words, in order to survive, it must germinate in spring, flower successfully in early summer and produce seeds. Accordingly, it requires a high input of nutrients, and since it lacks an established root network from which it can develop and grow, it grows nowhere near as tall as the common nettle over the course of the summer. There is much to suggest that the annual nettle only arrived in central Europe a few hundred years ago and thus does not belong to the native vegetation but is instead one of the 'old new plants', the archaeophytes. The common nettle most probably proliferated during the spreading of plant life that took place at the end of the last ice age, without the assistance of humans. If a plant is annual, this can signify transience, while being perennial can indicate permanence. Sometimes it is worth taking a more detailed view of the plants on which butterfly caterpillars live. Nettles are not all the same, even if their stings are equally unpleasant.

Other species of Lepidoptera live on nettles, among them two crambid moths that look quite distinct, but whose caterpillars are both as green as nettle leaves. These roll themselves up into leaf sacks, in which they live. Since these 'leafrolls' are rather conspicuous, it is useful at this point to draw a comparison between these crambid moths, the small magpie, *Arania hortulata*, and the mother-of-pearl, *Pleuroptya ruralis*, and those butterflies whose caterpillars also eat nettle leaves. Where the latter do this is very noticeable, since they normally huddle together in gregarious clusters and leave behind webbing, marked with caterpillar excrement, on the nettle plants. One might think that these clusters are not necessary because nettles are common and grow across large expanses. Nettles grow alongside nettles, full of life and vigour, yet most of them do not get eaten. Only in the areas immediately surrounding the caterpillar clusters are there visible holes in the leaves, the results of their hard work. If these aggregations of the same type of caterpillar appear repeatedly in similarly sized groups of plants, there must be a reason for it. The fact that there is an alternative is evident from the rolled-up nettle leaves, sticking out at right angles to the plant stems or inclined slightly downwards, which are occupied by the caterpillars of the mother-of-pearl moth. The fact that these caterpillars only carry a few thin and glassy bristles on their bodies, while the caterpillars of the peacock butterfly, the small tortoiseshell and the red admiral are almost bizarrely prickly, makes the clustering even stranger. The mother-of-pearl caterpillars clearly require the protection of the leaf roll. The peacock caterpillars are protected by their spiny coat. Or are they?

The caterpillars of all these butterflies ought actually to be comprehensively protected by the nettles themselves, whose pointed stinging hairs break at the slightest touch and inject irritating toxins into the skin of the invader. The most significant of these are acetylcholine and histamine, as well as some formic acid. Most people will know how much nettle stings hurt from their own experience. Why nettles (must) sting so sharply is a separate issue, which is also very interesting from a biological perspective. The answer, which we can provide with the greatest certainty, lies at the heart of the current crisis in agriculture, which is still a very long way from being adequately understood by the public.

Nettles are very rich in nitrogen and therefore highly nutritious plants, which proliferate wherever overfertilization occurs. Nettles require a high level of mineral nutrients, particularly mineral nitrogen compounds. Accordingly, they have always prospered in those places where, for whatever reason, there has been an accumulation of nitrogen (and phosphate) compounds. Historically, these were often resting places for larger grazing animals such as bison, *Bison bonasus*, and aurochs, *Bos primigenius*, in woods and on river plains. Their domesticated successors, domestic cattle, also rested in half-shaded, protected forest edges and near springs in pastures. The excretions of the creatures that gathered there fertilized the areas surrounding these resting places.

As is generally known, cattle digest grass and other food by rumination. What they have ingested fairly rapidly is regurgitated from the rumen and chewed again more thoroughly. They then swallow the greenish pulp again and it makes its way through the many-chambered stomach. The digestive system of ruminants is highly effective. Assisted by microbes, milk and meat are ultimately produced from grass or hay. The fact that a correspondingly large quantity of excrement in the form of soupy cow pats is accumulated in the process is also well known. When cattle are kept permanently indoors and manure is pumped from the stalls as slurry, the extremely unpleasant smell gets up our noses. When cattle were always outdoors, there was no slurry, but instead a complex and continual natural processing of the cow pats through a multiplicity of different insects, such as beetles and flies and any quantity of microbes. Without creating any stench, these 'dung processors' released the plant nutrients, thereby fertilizing the plants in these places. The nitrogen-hungry plants reacted to this with vigorous growth, above all the nettle and broad-leaved dock species, since neither of them were eaten by the cattle. Early German botanists grouped these plants together with the somewhat oblique expression 'cattle bed flora' ('*Lägerflora*'), meaning all those plants occurring where cattle rest.

Livestock farming has therefore benefited nettles since the time it was first practised. In areas with frosty winter weather, cattle had to be kept in stalls and pigs were penned in any event, which resulted in cattle bed flora developing around livestock stalls on farms as well, particularly where effluent runs flowed out of dung heaps. There, nettles grew rampant. They are, in effect, manure plants. It is easy to demonstrate that this classification is justified. One need only crush nettles slightly and place them in water. After a few days, the brew will be very pungent and produce an odour closely resembling liquid manure. Nettle tea is most suitable as a liquid fertilizer for increasing the growth of nutrient-needy plants in the garden. The process of leaving them in water simply releases from the nettles the nitrogen that they absorbed and converted into organic compounds during their own growth. Something very similar occurs with maize, which also grows exceedingly quickly if it is lavishly provided with nutrients. Under the right conditions, in just a few months, a tiny kernel of sweetcorn can grow into a plant more than three metres tall with a stalk as thick as a child's arm, and, in the case of grain maize, heavy, golden ripening corn cobs. I shall return to maize later, because it occupies a key position in the decline of butterflies, wildflowers and field birds. Here we are concerned with the fundamental biological connection between these two outwardly very different plants. Nettles and maize both have an extremely high nitrogen requirement, and a correspondingly high nitrogen content for herbivorous animals. Maize is also used for pig feed; in earlier times, nettles were often used as a vegetable and chopped up and given to geese and ducks, even chickens, as green fodder. Nettle salad is healthy. We just have to take care to 'deactivate' the acetylcholine and histamine that nettles contain when we prepare them. Small wonder that nettles are also attractive to the caterpillars of a very wide range of butterflies.

Nettles escape defoliation

Following this digression on the general connections, let us take a closer look at the special relationships between the nettle-feeding Lepidoptera. As emphasized above, these butterflies and moths should be extraordinarily abundant because there are so many nettles. However, if we look for nettle plants in gardens or forests, we will soon realize that it is not at all easy to find caterpillars on them. Indeed, we will discover a caterpillar web here and there, but the nettles will never have been defoliated extensively. Are the poisonous stinging hairs a successful deterrent against animal damage after all? In the case of cattle, goats or sheep, indeed they are. Roe and red deer avoid them too. However, the mucous membranes of herbivorous mammals react differently from the mouthparts of butterfly and moth caterpillars. Since they are specially adapted to digest nettles as part of their diet, one would have thought that they would multiply to such an extent that the area of nettles that they fed on would be stripped bare. But, to emphasize the point once again, this does not happen. We have not yet considered another species of butterfly that is relatively common in woods, and also has caterpillars that feed on nettles: the map butterfly, *Araschnia levana*. This is a rather special butterfly, since it comes in two forms: a light, brown-checked spring form and a dark, white-striped summer form. Only the upper side of the wing differs in this 'seasonal dimorphism', the technical term for this phenomenon. The underside is the same in both forms and bears a pattern reminiscent of a stylized map. From this we may conclude that it is advantageous for the map butterfly to be light brown-checked in spring but dark in summer, when the wings are open.

We will defer a closer inspection of this quirk until it comes up again in another context, and simply note that the caterpillars of the map butterfly only feed on the leaves of the common nettle. Now we already have four butterflies that rely completely or extensively on nettles as their larval host plant: the peacock, the small tortoiseshell, the map and the red admiral. To this we can add the two very common small moths from the crambid moth family that have already been mentioned (although, like the peacock, they can also fall back on various other species of larval host plant). I have already mentioned other Lepidoptera species whose caterpillars live on nettles, but their inclusion would make our relationships too complex and they have been left out for the purpose of our discussion. In any event, they do not change the decisive conclusion; if anything, they reinforce it: significant defoliation of nettles does not occur. The reason that this conclusion is so important that it merits repetition will become clear when we consider maize from the perspective of its possible insect pests.

Maize: damaged beyond repair

Its rapid growth, its cultivation over millions of hectares and its exceptional exploitability as feed for such a wide variety of animals should make maize at least as attractive to insects as nettles. But, unlike nettles, maize is threatened by huge numbers of pests, and is accordingly protected against them with large quantities of insecticide. Perhaps the difference lies in the content of the nettle's stinging hairs: the acetylcholine and histamine. Maize does not contain this type of chemical defence system.

There is another agricultural plant that belongs to our most important sources of nutrition, which does contain a chemical defence, namely the potato. Why does this crop need to be protected against the Colorado potato beetle, even though the surface (green) part of this plant from the nightshade family contains highly effective toxins, especially solanine? (Just 200–400 milligrams per kilogramme of bodyweight, that is, 20–30 grams for a person weighing 75 kilogrammes, can be fatal.) It is because specially adapted animal species, for instance the Colorado beetle, either convert the toxins into nontoxic substances through digestion or are not sensitive to them and are therefore able to monopolize these foodstuffs. For some agricultural crops, this can lead to serious infestations and to economic losses. These specially adapted species are considered pests and are, therefore, controlled. Yet nobody takes a chemical cudgel to the caterpillars of nettle-feeding Lepidoptera: on the contrary, nettles are disliked or classified as weeds, and their biological control by the caterpillars that feed on them would in fact be welcome. If herbicides were the only threat to nettles, and insecticides the only threat to the Lepidoptera that feed on them, we would, in theory, enjoy clouds of peacocks, velvety and carmine-banded red admirals, pretty and brightly checked small tortoiseshells and clouds of those other butterflies – with, in the spring generation, wings resembling the small tortoiseshell – the charming map butterflies. Bats could catch the slow-flying mother-of-pearl and small magpie moths in the dusk and early evening hours. But nothing of the sort happens. Quite different Lepidoptera species are responsible for defoliation and mass proliferation. The combined use of nettles by the caterpillars of different, quite numerous butterflies does not destroy the nettles. Even if the nettles were still used commercially for the extraction of nettle fibres, these butterflies would not be considered pests, and never have been.

Why is this? The butterfly collectors of earlier times knew the reason, since it affects the vast majority of butterflies. It is this: caterpillars are normally subject to frequent attacks by parasites. The losses caused by parasitic insects such as the (tiny) ichneumon wasps, the small Braconid wasps and the larger tachinid flies maintain the populations of nettle-loving butterflies and moths well under the damage threshold, that is, below a level of infestation that would be conspicuous to us. These insects that live inside the bodies of the caterpillars in their larval stage are known as parasitoids. They should not be confused with external parasites such as fleas and lice.

Cabbage whites: parasites and protection

Infestation by parasites once curbed the populations of cabbage white butterflies too. We always had large white caterpillars at home in the garden, on the different kinds of cabbage that we grew. The damage that they caused with their feeding nevertheless remained minimal – surprisingly minimal, since we used no poison at all to protect our plants in the 1960s. The fields of green cabbage and also those of red cabbage were not sprayed either. We were able to rent a few rows of a field from a farmer in the village in exchange for our help with the cabbage harvest. We had to look after these ourselves, in order to produce healthy cabbages. This just meant weeding; apart from that, they looked after themselves. One put up with the fact that the hares would eat a few of them in the autumn, before the harvest. There were many hares, as evidenced by the hunts in late autumn. Whole wagonloads of hares and pheasants would be driven into the village after each half day hunt.

Nobody complained about hare or caterpillar damage to cabbages. In our garden, the caterpillars of the large white that were ready to pupate would crawl up the house walls. Soon most of them would be covered with dozens of yellowish cocoons, which people referred to as 'caterpillar eggs', since they were so common. Naturally there were no eggs involved. They were the cocoons in which the larvae of a parasitic Braconid mini wasp, *Cotesia glomerata* (formerly *Apanteles glomeratus*), that fed on cabbage white caterpillars had pupated. The majority of cabbage white caterpillars did not even achieve pupation at that time. Those that did so were by no means all successful. Often, they did not produce butterflies. I learnt this through my first attempts to 'breed' butterfly caterpillars. After a while, most of the pupae were so full of holes that they looked as if they had been attacked with a miniature shotgun. The tiny larvae of the chalcid wasps, *Pteromalus puparum*, had eaten them from the inside out while the butterfly was supposed to be developing. The wasps themselves emerged from the small holes, leaving nothing but an empty husk. A cabbage white only emerged from a few of the pupae. Just like a lottery, most of the caterpillars and pupae yielded only 'no wins'.

That is how it was in the late 1950s and 1960s. These days, the situation is different. It can be quite difficult to grow kohlrabi or other types of cabbage successfully in the garden, because they are eaten by caterpillars. Yet cabbage whites have not become more abundant; in fact, the opposite is true, as I shall explain in connection with the migratory flights of cabbage whites. Perhaps there is a shortage of food plants growing in the wild, and so they come to our gardens to lay their eggs. The females that are ready to lay eggs are never significantly disturbed by our presence. Anyone who sees us groping around in the vegetable patch with our arms, or a long stick, may get the impression that we are trying to drive out the cabbage whites with some kind of magical incantation. But of course, without success.

The blackbirds that scratch around in the earth next to the plants also do little to combat the caterpillars. They are interested in the earthworms that are now in such good supply thanks to the composting of biodegradable waste. Birds do not like large cabbage whites at all. They taste unpleasant because they have absorbed the toxins and other protective substances from the cabbage plants. Their caterpillars ingest them and pass them on to the butterflies via the pupae. They are poisonous glucosinolates. Thanks to these 'weapons', the leisurely flight and the conspicuous markings of these white butterflies do not put them in danger.

Peacocks, small tortoiseshells, red admirals and map butterflies do not contain any similar protection against predators. The nettles that their caterpillars feed on provide them with nothing that would help. Yet they are neither noticeably rarer than the cabbage whites nor particularly endangered by the behaviour of birds. Even if we looked at them more closely when they visit our gardens, we would barely notice that they have a few triangular beak marks in their wings. Their erratic flight evidently provides sufficient protection from hungry birds. For this reason, we need not worry that greater protection of our songbirds will have a negative impact on the butterflies in our gardens; certainly not on the Nymphalid butterflies that are so flamboyant and universally loved.

The mass flight of the map butterfly: singularities in the realm of the butterflies

My digressions to map butterflies and the cabbage whites seem to have contributed little to the question of why nettle-loving butterflies are not much more abundant, given that nettles grow in large quantities almost everywhere. But this is typical of the search for explanations for processes in nature. It is rare to arrive at a convincing result in a swift and direct manner. Often a side note provides help, or a small observation points the way forward – for example, information from an amateur group. Sometimes, something will occur to a butterfly-lover, either a collector or a researcher, that initially seems irrelevant. Thus, in the large and excellent 10-volume handbook, *Die Schmetterlinge Baden-Württembergs* [*The Butterflies and Moths of Baden-Württemberg*], edited by Günter Ebert (Staatliches Museum für Naturkunde Karlsruhe, volume I, 1991, page 405), we read about map butterflies: 'occurrence and regional distribution of this species are clearly influenced by sporadic but extreme population fluctuations'. Schneider (1936) indicates frequencies of occurrence that change from year to year, and notes that, 'in the vicinity of Stuttgart, [they have been] very rare for many years, but growing more frequent in the last few years'. Lindner (1935) explains this as follows: '*Araschnia levana* and *prorsa* had almost disappeared from Württemberg for many decades. The oldest collectors could only recall occasional finds going back many years. This made the population explosion of the insect in the whole country in 1932 all the more surprising.' A year later this was followed by an increase in tachinid flies, which 'caused an immediate fall in the butterfly numbers'. These details provided by amateurs do contain part of the explanation: the shortage of parasites that led to the population explosion of 1932. However, the summer of 1932 was, according to meteorological data, a completely normal, average summer. The mean temperatures of the previous and subsequent summers were also within the normal range. Consequently, the natural main enemy, the parasite of the caterpillar, must somehow have been drastically reduced prior to the mass increase.

The same was true, in all probability, of the population explosion of map butterflies that I experienced and documented in 2013 (see Photo 10). At the time, I counted more than 1,300 of these butterflies on a single inspection round in the commercial forest in midsummer. They were so numerous that they jostled with one another to find a spot on the blossoms at the edges of the forestry tracks. Some plants were completely shrouded in map butterflies. I even found one that had got caught on a burr. In the riparian woods they were not quite as abundant, but were present in far higher numbers than average. They flew to places where I normally never saw them, and even visited gardens. The number of map butterflies observed on a single day was far in excess of the total count for a whole decade.

Thanks to exact counts and temperature recordings, the background to the population explosion of 2013 can be reliably reconstructed. In February 2012, there was an extreme period of frost. For two weeks, the nightly minimums fell below minus 20°C. There were also regular daytime frosts. The cold arrived suddenly after a December and January that had been too warm. The (dry) cold clearly did not harm the hibernating pupae of the map butterflies. This was borne out by the butterfly counts that were undertaken in the spring. The following generation in summer 2012 was significantly larger, but was not yet really remarkable. However, it evidently produced far more hibernating, parasite-free pupae than normal because, in spring 2013, there were already an abnormally high number of map butterflies.

This was surprising, since during their main flight time in May it was considerably too cool and, above all, very rainy. At the end of May/beginning of June 2013, the persistent bad weather led to the 'flood of the century' in the northern pre-alpine regions. The riparian woods along the River Inn were extensively flooded. Small wonder then that noticeably fewer map butterflies flew in these woods than in the commercial woodland that had not been affected by the flood. For the parasites, the extreme frosts of the preceding winter and the bad weather in the first half of 2013, including a frosty late winter that lasted well into April, were evidently extremely unfavourable. The map butterflies escaped their usual decimation by the parasites. And as we had persistently hot, dry weather from July onwards, to compensate for the weather in the first half of the year that had been far too wet and cold, conditions were ideal for the map butterflies. They flew in great numbers 'as never before'.

The note in the butterfly handbook referred to above turns out to be the key to the question of why there are sometimes population explosions of butterfly species whose caterpillars feed on nettles. Unusual weather conditions, which perhaps did not suit the parasites, could be verified. Moreover, I had already experienced a relatively large population of map butterflies in spring 1986 in the floodplains of the River Inn, and recorded counts.

In 1984 and 1985, according to the values measured at the Hohenpeissenberg Weather Station south of Munich, the summers were unremarkable and average. However, the average temperatures for the winter prior to the mass flight years were remarkably similar, at minus 3.1°C, minus 2.9°C and minus 3°C. They indicated cold, but not extreme winters, with pronounced frost periods. The winter prior to the notable increase in map butterflies in Baden-Württemberg in the 1930s was the very cold one of 1929/30. With an average of minus 5.7°C, it was the second coldest winter recorded in the whole of the twentieth century. Only the bitter winter of 1962/63 was colder, with an average of minus 6°C. Before this, there were only two equally cold winters in the whole of the nineteenth century (each minus 6.1°C), those of 1829/30 and 1894/95, and one of minus 5.5°C in 1890/91. In Britain, temperatures were so low that the Thames froze in 1829/30 and the winter of 1894/95 was also very cold. The inference was clear: extreme winter weather had reduced the parasite population to such an extent that the populations of map butterflies were no longer held in check by them and the butterflies suddenly became extraordinarily numerous. But their escape was short-lived. Already after two, at most three, generations of butterflies, the parasites caught up with them again, and forced the maps back to their normal low levels. The map butterfly population in 2014 was also no different from the populations recorded prior to the population explosion the year before. There have been no further such occurrences since then. It is not possible to predict them, since they depend not only on special environmental circumstances; the population itself must also be in the right condition.

This means that population explosions must be excluded from trend analyses, since, as isolated cases ('singularities'), they would influence the statistics too strongly. For specific purposes, however, it is appropriate to look at these instances in isolation: how frequent they are in given time periods and how often they occur. In the first third of the twentieth century the population explosions of forest insects were strikingly common and also very large. Whole forests were destroyed because the pine tree lappet, *Dendrolimus pini*, the pine beauty, *Panolis flammea*, and even the larger pine hawk-moth, *Hyloicus pinastri* (as well as the bordered white, *Bupalus piniaria*, and the pine beauty in Britain) fed so intensively on coniferous forests. What triggered these mass population increases remained unclear, despite investigations by forestry experts. Most probably the trees just happened to be the right age for the caterpillars to feed on. Moreover, these were single tree species forests with little genetic diversity. This lack of genetic diversity generally differentiates the vast majority of crops from natural populations of the same plant species.

The population explosions of forest pests mentioned above were incomparably greater than that of the map butterfly in summer 2013. But high figures can be misleading when they are not considered in the context of the available nutrition: 1,300 map butterflies a day seems a strikingly high number when compared to normal figures for the summer flight, which had usually produced 10–15 specimens on transects of the same length. In other words, the population had increased by a factor of roughly 100. But what did this mean with respect to the nettles? They showed, as I have repeatedly emphasized, no noteworthy degree of visible damage. Why this should be the case becomes immediately clear when the nettles are counted for comparison. With hundreds of plants per square metre along the forestry tracks, there would have been hundreds of thousands along the 4-kilometre transect. The caterpillars therefore consumed nettle leaves in proportions per thousand. From this, it is evident that even in the case of these population explosions, the consumption of nettles by caterpillars has no effect whatsoever on the numbers of nettles.

Does climate change affect the seasonal morphs of the map butterfly?

Two further peculiarities of map butterflies should be considered. As explained above, this butterfly occurs in two forms, a light yellow-brown, black-chequered spring form, f. *levana*, and a summer form, f. *prorsa*, in which the upperside is dark with a broad white band over the wing (see Photos 11 and 12). The two forms always fly separately. The dark summer form develops from caterpillars that are exposed to the long and lengthening days of early summer. The black-chequered spring form develops from caterpillars that are alive during the shortening days of late summer and autumn. The length of the day works like a timer. The same must be true for all other butterflies that have an early and a late summer generation of caterpillars. But other than the map butterfly, none of them develops such a clear, pronounced seasonal dimorphism. From this we can assume that it must be especially important for the maps to have a different appearance in the main flight time in spring, in April and May, to later in the summer.

It is so important that there are no deviations from the seasonal differentiation. To pre-empt the question: we do not really know the reason, but we can make educated guesses. For example, we know that we find it much harder to see the light-brown chequered spring butterflies in the relatively bright light that finds its way into the deciduous forest at this time of year. The same is true for the summer form in the stark contrasts of light and dark that are typical of a forest in high summer. Since they rely on contrasts and shape recognition, birds might have similar problems seeing these butterflies. Moreover, there are numerous butterflies, particularly in the open landscape, that have a bright, spotted pattern on the upper side of their wings. In high summer, it resembles the black, white-striped form of the significantly larger white admiral, which flies in similar places to the map butterfly. As suggestive as this implied similarity might seem, it is not sufficient to explain the strict seasonal dimorphism. For this reason, the fact that map butterflies are weak and not very agile flyers is presumably also relevant, together with the fact that they carry no toxins or defensive substances in their bodies, since their caterpillars fed on nettles, which will have provided no protective substances. The predation pressure of birds on the butterfly stage of the maps ought therefore to have a similar effect to parasitization on the caterpillar stage. Butterflies and caterpillars inhabit different environments and are therefore exposed to different effects from weather and predators.

The weather is relevant to the second peculiarity that should be addressed. If there is an early spring and a warm early summer, the flight of the midsummer generation of maps will start correspondingly early. In 2018, I had already seen the first dark one (of the second generation) on 9 June. Just how early this is, is apparent from the handbook *Die Schmetterlinge Baden-Württembergs* cited above (volume I, page 406): 'In very hot years, butterflies from the summer generation were already caught at the end of June (24.06.64, 30.06.46)'; and 'Some individuals from the first generation flew until the middle of June.'

Is such an early appearance of the summer generation a sign of the current climate change? Such scattered data at the beginning and again at the end of the normal flight time would normally prove nothing except that the weather happened to be favourable. However, with the map butterfly it is different. If the summer generation starts its flight very early, then another generation develops in late summer. The butterflies from this later generation, dark like those of early summer, emerge from their pupae in mid- to late August. They are then visible until the beginning of September. Indeed, there has been such a second generation in summer in recent years, for example at the end of August 2012, but whether it contributed to the mass flight of 2013 is unfortunately not a question that we can answer retrospectively. It is possible that they could have escaped the pressure of the parasites through their untimely appearance.

Is this therefore a case that points to climate change and its consequences? Older findings indicate that scepticism is in order. The handbook cited above, *Die Schmetterlinge Baden-Württembergs* (volume I, page 407) reads: 'Certain butterflies ("map butterflies") from the 3rd generation were noted in hot years from 20 August' – unfortunately the year is not specified. My own notes from the Lower Bavarian Inn Valley, whose climate is cooler than that of the Upper Rhine Valley, contain several references to third generation map butterflies from the end of August/start of September between the 1960s and the 1980s. Their occurrence depends, therefore, simply on the weather experienced in spring and early summer, and not on the midsummer temperatures that are increasing by tenths of a degree.

Nature is too diverse for simple generalizations

On the basis of the findings mentioned above, the geographical distribution of the map butterfly is easy to understand. It extends from the Pyrenees in a wide arc over France and Germany, towards central eastern and eastern Europe. On the Iberian Peninsula, in Britain, in France in the Mediterranean south and in the mild Atlantic northwest, in the whole of the Mediterranean as well as in some regions of northern Germany, map butterflies are absent or scarce. In other words, the species does not occur in regions that have mild winters, in which there are no pronounced, long-lasting cold periods. The note for Baden-Württemberg in Ebert's book cited above fits this observation too. So, is the map butterfly an exception to the rule that most butterflies prefer mild winters without long periods of cold? Caution is advisable when making generalizations. It is all too easy to run the risk of giving excessive significance to individual findings. However, what we have seen here fits together like a mosaic to create a result of much greater consequence.

Not only the map butterfly, but all the nettle-loving butterflies are subjected to pressure from parasites. We can see this directly from the behaviour of the caterpillars. Congregating in clusters reduces the probability of individual peacock caterpillars being stung by an ichneumon wasp and parasitized. This is because the more neighbours a caterpillar has, the smaller the chance that it will be affected. The fact that the mother-of-pearl caterpillars roll the leaves up into pods that seem too spacious for them, and of which many are empty, points in the same direction. Ichneumon wasps and tachinid flies cannot find them so easily within their leaf-rolls, especially when many of those they do find are empty.* Silken threads at the ends of the rolls can warn the caterpillars of danger when they are touched by parasitic wasps. Sensing this movement, they will fall out of the rolls with lightning speed and land on the ground where they are almost impossible to find. The nets built by the small caterpillars of the peacock and the small tortoiseshell also offer a certain amount of protection. Their significance will become abundantly clear when we discuss small ermine moths. Caterpillars do not simply spin for their own pleasure. They could save the chemical raw material for the thread and use it for other purposes if tent-like nets

---

* It is worth noting that there is a specific Braconid parasitoid that thrives on these leaf-rolls, using chemicals on the caterpillar silk to distinguish between old (and therefore empty) and new (occupied) leaf rolls.

were not important for their survival, but the matter of eating or being eaten in the insect world is far more dangerous than we sometimes appreciate. We are simply not able to follow the events surrounding the butterflies and moths any more closely.

Mostly, we only see the result – that is, the fact that the activity of the caterpillars of the nettle-feeding Lepidoptera does not reduce the quantity of nettles to a degree that would be conspicuous. 'Peacock & co.' are therefore not likely to be employed as a biological agent against nettles (see Photo 13). It is thus all the more surprising, even unnatural, that the insects that live off cultivated plants need to be so strictly controlled.

The counts of peacocks and other such species carried out over many years have resulted in a few further quite unexpected findings. The populations of nettle-loving butterflies vary from year to year. Their populations fluctuate much more widely than they ought to, since the nitrogen surpluses delivered by agriculture have persistently favoured the growth of nettles for decades. If the parasites were actually to compensate for the improved living conditions of the nettle-loving butterflies, they would need to have a similar effect on the different species. But while the peacock butterfly and the comma, *Polygonia c-album*, another nettle-loving butterfly, have become significantly more abundant recently, the small tortoiseshell has become rarer most years. For some years I have hardly any records of it at all.

Word of the decline in small tortoiseshells may not yet have reached some nature conservationists and butterfly lovers, although conservationists in the United Kingdom may be aware of it following the arrival in 1998 of a parasitic tachinid fly, *Sturmia bella*, previously unknown there. On the whole, the species is considered to be as plentiful as ever. It is, but only in single years, and no longer every year. Conclusive results are not completely overturned when an example appears to prove the opposite, but it does test their limits. Nature is too diverse for simple generalizations. Nonetheless, the small tortoiseshell and the wide fluctuations of the peacock lead us to a phenomenon that is still barely understood: the migrations of butterflies and their significance for species conservation. This in turn shines the spotlight on climate change, which these days seems to serve as a universal explanation for any changes in nature that are not fully understood or that have yet to be researched in any detail.

The Great Migrations of the Butterflies

I found an extremely unusual entry in my old notes on butterflies and moths. On 9 March 1959, a 'wonderful spring day', as I noted, I recorded a painted lady, *Vanessa cardui*, as the first butterfly of the year. I was on my way home from school at lunchtime when it flew across the road directly in front of me. This is an extremely early date for the painted lady, which usually migrates to Germany from the south much later in the season. It is weeks before the date of 1 April 1966, recorded as the earliest confirmed sighting in the handbook *Die Schmetterlinge Baden-Württembergs*. At 3.20pm I went off to the riparian woods along the River Inn and saw, as I crossed the meadow, 'several small tortoiseshells', as I remarked at the time. 'The butterflies were sunning themselves and there was a southeast wind.' This type of wind usually occurs in the Lower Bavarian Inn Valley when a *föhn** blows from the Alps, far into the pre-alpine region. For the painted lady, as for the small tortoiseshells, there was therefore a gusty wind blowing from the south that day. I noted 16°C as the maximum air temperature. The previous year, 1958, was a year of 'high numbers' for the painted

---

* A *föhn* (or *foehn*) is a type of dry, warm, down-slope wind that occurs in the lee of a mountain range. A *föhn* wind can raise temperatures by up to 14°C in a few hours and is usually associated with very good visibility.

lady in Baden-Württemberg, according to the handbook cited above. It was highly unlikely that one of these had hibernated in our region. The winter of 1958/59 had been much too cold for this. According to the measurements taken at Hohenpeissenberg, it lay within the range for a normal winter. In January 1959, I had recorded minus 15°C outside the house with my maximum–minimum thermometer.

No painted lady can survive such cold, not even in relatively protected areas such as those the peacocks seek out for hibernation. The painted lady that I saw in March had therefore, with relative certainty, flown over the Alps shortly beforehand. Most probably it was carried along by the *föhn*. It is quite normal for painted ladies to fly from Africa as far as Italy or even to England in March, and to produce the first generation of the year there. The first one that I saw in southeast Bavaria on 19 April 2018 was surely such an early visitor. The preceding winter had been far too cold for a successful hibernation in the foothills of the Bavarian Alps. On 28 February, the minimum night-time temperature in the region was minus 13°C and the maximum daytime temperature was only minus 6°C.

The migratory flights of the painted ladies

We now know that the migrations of the painted ladies extend from the Sahel region, the southern transition zone of the Sahara Desert in the north African savannah, over southern and central Europe, and northwards up to north, north-western and north-eastern Europe. In mid- to late summer they endeavour to return to Africa. The return flight is often made using an easterly route over the Bosphorus, western Turkey and Lebanon. In this manner, assuming the weather that year is overall favourable, they end up travelling in a huge circle, 5,000 kilometres in diameter. The whole flight, though in fact a sequence of shorter flights by several generations, is more than 15,000 kilometres. These are distances similar to the furthest covered by migratory birds.

In spring, the painted ladies use south-westerly and southerly breezes and travel northwards from the western half of North Africa with their aid. In late summer and autumn, the butterflies of the generation bred in northern Europe fly back in the direction of Africa with winds that blow from the northeast and the north (see Photo 14). In the meantime, they produce two generations of new butterflies. But this is a highly simplified explanation. Year on year, each generation of painted ladies adapts itself to the actual circumstances of precipitation and wind. In some years they do not come over the Alps to central Europe at all, or only singly. In a few, special years, however, they stream north and northeast in their millions, just like in spring 2003 and again in 2009. More of this below.

We must not lose sight of the small tortoiseshells that I noted on that 9 March 1959. We are actually quite familiar with the transition between permanent residence and regular migration for small tortoiseshells. They are, as it were, the partial migrants among the butterflies, comparable to some of our bird species, in which a part of the population spends the winter here, and the other, usually the larger part, leaves for the south and overwinters around the Mediterranean, returning the following spring. The small tortoiseshells of 9 March 1959 were just such migrants. Now, almost 60 years later, I can assert this with great certainty. Once my attention had been drawn to it, I recognized their spring migration every year. But they always came in very variable numbers from the south. The river valleys stretching north to the Danube act as direction markers for them. This finding will, when considered more closely, provide an explanation for why the numbers of small tortoiseshells differ more from year to year than the peacocks and other butterflies whose caterpillars live on the abundant nettles.

Small tortoiseshells as travellers

Let us imagine, therefore, an (early) spring day in the Isar Valley to the south of Munich, on which the *föhn* has caused the sky over Upper Bavaria to turn an azure shade of blue. Thin white clouds drift towards the northeast. The warm wind surges through the valley. Close to the mountains, the wind can be so strong that it is called a '*föhn* storm'. If it hits you head-on, you have to brace yourself or you will be knocked down. As a rule, butterflies avoid any strong wind. It moves them along uncontrollably and prevents them from looking for flowers containing nectar or mates with which to pair. And yet it is just such conditions in which we can see small tortoiseshells rapidly streaming northwards with the wind, focused on their goal. They hurry along, as if pulled by an invisible string. Now and then one of them will land on the ground and, if the air is warm enough, briefly stop and warm itself with outstretched wings.

It can be 20°C or more when the early spring sun shines brightly at the end of February or the beginning of March. On such days the butterflies will typically fly at a height of 1–2 metres above the ground, in quite a straight line, in contrast to their usual erratic movements when they are looking for flowers. Sometimes they pass people just at the last minute. If a small tortoiseshell comes close to your face when it is travelling at such speed, it can be difficult to recognize. In moments such as these, one sees how it is anything but easy for a bird to catch a butterfly that is flying towards it, even though the bird can fly much faster. If a single small tortoiseshell does fly past at speed, it is worth settling down and observing what happens next. Soon it will become evident that many small tortoiseshells are hurrying past, one after another. They are all travelling in the same direction: mainly northwards, but always along the river valley. There might be a dozen or more per hour, especially between 11am and 3pm, the main flight time. Sometimes they come one after the other so fast that a quarter of an hour is enough to convince oneself that the spring migration of the small tortoiseshells is taking place. But when the wind blows from the west or the northwest, even if it is not a particularly strong *föhn*, and if the clouds only part occasionally to let the sun through, we will wait for them in vain.

The weather conditions need to be just right. If there is pronounced westerly weather without any southern currents or *föhn*, as is so often the case between the end of February and mid- to late March, then the migration of the small tortoiseshells will probably not take place. They will then not be present during the summer months, or only rarely, depending on how many managed to hibernate as butterflies in the northern foothills of the Alps. It is almost impossible to determine what proportion of the butterflies do this, since one cannot check every nook and cranny into which the small tortoiseshells might have disappeared to hibernate during the autumn. Most of them are sighted when they get into trouble, flapping at the windows of greenhouses or garden sheds. In March or early April, they seek the light, but they cannot cope with glass. In autumn they will have looked for dark cracks and porches, since their internal behaviour pattern is tuned to natural hiding places such as caves or rock fissures. The fidgety small tortoiseshells – and peacocks, which behave quite similarly – might give the impression that many of them overwinter with us. But when I compare my own findings of hibernators with the counts of incoming butterflies, the difference in quantity is very clear. It is possibly a small percentage of the autumn population that hibernates, but by no means all survive: many perish because their hiding places are too warm, for example in houses, or too cold, for example in places that they have found outdoors. This assessment is supported by the fact that the numbers of small tortoiseshells fluctuate significantly from year to year. Those that hibernate successfully clearly cannot compensate for the lack of migration if weather conditions over the Alps are unfavourable. Observations in the high mountain ranges of Switzerland revealed many years ago that small tortoiseshells assemble in great numbers on the southern side, clearly waiting for favourable weather to make the onward flight over the alpine passes (see Photo 15).

They certainly could not have spent the winter up there at the treeline. We do not yet know enough detail about how this seasonal migration of small tortoiseshells or the geographically far larger one of the painted ladies is played out, but the broad outlines are now becoming apparent, in the same way as they did for the migration of birds 100 years ago. At the time, the first tagging of birds with tiny aluminium rings ushered in the era of scientific bird migration research. Thanks to the size of many birds, their migrations can now even be followed individually. We can see on the computer, for example, where the greater spotted eagle Tönnes* is right now, exactly where he will migrate to in Africa, the winter home of the painted lady, when he starts his journey back from Spain to his Baltic home, and which loops and deviations from the main route he makes on the way. The miniature technology that would enable this to be done with small tortoiseshells or painted ladies has not yet been developed. However, detailed chemical analyses already allow us to confirm which part of Africa the painted ladies who migrate to central Europe in spring come from, since they carry the

---

* An individual eagle from Estonia whose GPS coordinates are plotted on various websites.

signature of the region in their bodies, through the nutrition that was consumed by their caterpillars.

Modern science and the private commitment of lay people – the citizen scientists – driven by their enthusiasm for butterflies, come together in a particularly intensive and fruitful manner in the investigation of butterfly migrations and, moreover, in the documentation of the incidence and fluctuations in populations of the different species. More of this later, since it is of central importance in understanding the factors that have led to the sharp decline in butterflies. The migratory butterflies regularly seem to belie this decline with their mass flights from the south, giving the impression that things might not be going so badly for our butterflies at all. They sometimes gather in such quantities on the flowers of a butterfly bush that its branches threaten to break under the weight.

Butterfly invasions

Let us return to the painted lady and its huge migrations. The two largest migrations in the twenty-first century so far took place in central Europe in 2003 and in 2009. According to estimates, which are naturally very approximate, since the network of observers was too small and no objective methods were available to record the quantities of butterflies that flew past, there were 20 million or more painted ladies on the wing on each occasion. However, there could have been 10 times as many. An estimated 11 million painted ladies arrived in Britain in the 2009 invasion. As is generally known, large numbers are impressive due to their sheer scale rather than because of any easily verifiable comparisons. Yet the impression created by direct experience can be even more lasting, such as watching the gathering in late afternoons and evenings of resting painted ladies in the Munich Botanical Gardens in 2003. Their numbers were so vast on some bushes that they weighed down the branches. The bushes carried far more butterflies than leaves.

Their imminent arrival had been indicated by the influx of a kind of advance party. The first painted ladies arrived at the start of May. On 5 May 2003, one of them flew over the Bavarian State Collection of Zoology (ZSM) in Munich at 7pm, heading north. The day had been hot; the maximum temperature was 30°C, the heat being caused by

Saharan air, which pushed the temperature up to 33°C the following day. On that day I saw two painted ladies fly over the ZSM in the early afternoon. The first butterflies thus arrived quite early in the year. This was not particularly remarkable, since the spring had been very warm, which suited the butterflies. The bright peacocks were more notable, since they were not in nearly as much haste as the ochre-coloured painted ladies, which barely settled anywhere and, in their purposeful flight, swiftly disappeared from view. After the middle of May, which brought a short cold snap, observations of painted ladies flying north increased. A few were seen almost every day, but since most of them were 'dawdling', they did not seem like the vanguard of the event that took place quite suddenly on 1 June, following a very hot end to the month of May.

I first saw a painted lady in the morning. Shortly after lunch came four more, which were clearly in a hurry. Things really got going in the afternoon. In the course of 15 minutes I counted 13, then more than 50, 100, then 175 from 5.51pm to 5.55pm, and more than 500 between 6.15pm and 6.45pm. I recorded the last one at 9pm. The migration continued on 2 and 3 June. The flight always reached its maximum intensity in the late afternoon between 5 and 6pm. The hourly distribution of the painted ladies and their flight speed, which, depending on the strength of the following wind, reached 30–40 kilometres per hour, suggested that the butterflies had started that morning at the southern edge of the Alps, having let the morning sun warm them before setting off. After flying for 8–10 hours, they arrived in the Munich area. The butterflies largely stuck to the Isar Valley, then cut across Munich, always maintaining a north by north-easterly direction, even when higher buildings were in their way. They flew over these without altering their course. Similar behaviour was observed by a colleague on 28 May 2009 during a northerly migration of painted ladies over Bristol in southwest England. After flying up and over tall trees at an average height of 30 metres, they descended abruptly before rising again to fly over the next obstacle, a three-storey apartment block.

In Munich, thousands arrived in the concourse of the main railway station, where most of them perished, since they could not change their direction of flight and fly back. Only a few were killed by cars, since the butterflies crossed the motorways near Munich almost without exception at a safe height of 2–2.5 metres. There may have been some collisions with lorries. In the Isar Valley and in the open fields outside Munich, they flew at a lower level, so that painted ladies touched our faces with their wings more than once, only partly avoiding us. In order to get an idea of the width of the migration, I drove west as far as the Lech Valley near Augsburg, and a similar distance of about 50 kilometres east. Friends and colleagues provided reports from river valleys even further west and from the Inn Valley.

From the transect widths and the numbers recorded per 15 minutes, we were able to calculate approximate figures for the total number of painted ladies. There were, without doubt, millions, many millions. There must have been more than 20 million that flew through the Isar Valley alone. A similar order of magnitude was estimated for the flight that took place in early summer 2009 in Austria, further to the east. At that time, they came mostly over the Salzach Valley and the Alpine valleys and passes further to the east. Few remained behind in the northern Alpine foothills. In the summer of 2003, that legendary 'scorcher', with Mediterranean weather from April through to September, painted ladies fluttered about in July and August together with native butterflies, feeding on nectar-rich blooms – above all on the butterfly bushes – but it was rare to find any caterpillars. Without doubt, the main bulk had flown further north and northeast up to the Baltic and to northwest Russia. At the time, the lower pressure centres also moved this way and brought plenty of summer precipitation to these regions. Further south, from France across Germany and towards Poland and the Czech Republic, they were absent.

The few summer storms, despite delivering large amounts of rain, did not remedy the precipitation deficit, so the summer of 2003 was not only very hot but also very dry. In the Munich region, many trees started to lose their leaves as early as the beginning of August. For the painted ladies it would not have been worthwhile to end their long-distance flight over the Alps in an area that was subject to the same heat-stress as the regions around the Mediterranean. Considering the event in retrospect, one could therefore be forgiven for thinking that the painted ladies 'knew' at the start of June 2003 that such an extraordinarily hot summer would develop to the north of the Alps. But a much simpler explanation is this: they flew to those regions where the spring weather fronts brought plentiful rain to the Atlantic lower pressure centres, and where ground vegetation thrived accordingly. In summer 2009, this was largely already the case in central Europe; as a result, many painted ladies laid their eggs here and, following the successful growth of their caterpillars and emergence from the pupae, their journey south took place in midsummer. More painted ladies flew south from here in summer 2006, following a less conspicuous migration that spring. This return flight lasted almost two weeks and was easy to observe since the flight speed of the painted ladies was quite slow. They stopped to 'refuel' repeatedly on flowers, but then flew in quite straight lines towards the south and southwest, without spending any longer in just one location, such as a garden.

Migrations of painted ladies have surely been going on for a long time. Recordings of them go back centuries. In the handbook *Die Schmetterlinge Baden-Württembergs*, large migrations are recorded for the years 1879, 1918 and 1931. A mass flight with 'hundreds of millions of butterflies' in southern France reported by J. Thiele (and cited in the above-mentioned book) only touched southwest Germany, although there were very large immigrations into Britain in 1879, 1903, 1928, 1934, 1943, 1945, 1947, 1948, 1949 and 1952. There was also a mass migration through Austria in 2009, with huge numbers of butterflies arriving in Britain that year from March to May.[*]

This type of natural phenomenon demonstrates very clearly what otherwise remains largely hidden – that is, the significance of suitable conditions for insect reproduction. Butterflies do need flowers that provide them with nectar and thus with energy, but it is the food plants of the caterpillars that determine their occurrence and frequency. Painted ladies are not especially selective in this respect. Their caterpillars can develop on a multitude of plant species, not just the thistle from which their German name is derived (*Distelfalter* = 'thistle butterfly'). Thistles are important, provided that they are growing and not yet dried out (by the summer heat), but burdock, nettles, coltsfoot and other quite different plant species also make suitable food for their caterpillars. There are also African plants that feed the winter generation. This means that

---

[*] See S. P. Clancy, 'The immigration of Lepidoptera to the British Isles in 2009', *Entomologist's Record and Journal of Variation* 124 (2012): 105–139.

the state of the vegetation is at least as important as the plant species themselves. We will return to this when we discuss the effects of excessive fertilization on butterflies and other insects.

Migratory butterflies, such as the painted lady, make it clear with their searches and their long-distance flights how the lifecycles of butterflies have evolved. Often, we will only see small fragments or parts of these lifecycles and interpret these as being particularly significant. They might be, but need not be. The arrival of the small tortoiseshells, which imbibe nectar from the floral clusters of *Buddleia* in mid- or late summer, discloses little about their origins or how successful the past breeding season was for them. It may have been favourable locally, and the butterflies might therefore come from that same garden or from its immediate surroundings. Then again, one might be looking at migrants that come from far away and are on their passage south for hibernation.

If the painted ladies do briefly inundate the landscape in their millions, they should on no account be included in the calculations of native butterfly populations: that would completely change the statistics and produce a false impression. This is true not only for painted ladies. There are other migratory butterflies and moths that fly here from the south in comparable numbers, but are less conspicuous because they are small, grey and largely active at night. For example, the silver Y moth, *Autographa gamma*, which belongs to the Noctuid moth family. Its characteristic mark is a silver letter 'y' that resembles the handwritten (Greek) letter gamma, in the middle of the forewing (see Photo 16). In midsummer and late summer, silver Y moths like to visit the flowers of the butterfly bush even during the day. During the UEFA European Championship in France in July 2016, they made themselves conspicuous and 'caused a public nuisance' as they flew into the stadium in large numbers and irritated the players. Pictures of the famous referee Pierluigi Collina attempting to defend himself against them went viral.

Silver Y moths fly over the Alps towards central and northern Europe every year. In mid- and late summer, and sometimes far into the autumn, their descendants return home because they would not otherwise survive the winter in significant numbers. The numbers involved in this moth migration are definitely in the millions in certain years. They probably outnumber the painted ladies several times over, but we notice too few of their migratory flights. The large, day-flying painted ladies are simply more conspicuous.

It is therefore useful to compare the painted lady to the small tortoise-shell that belongs to the same family group (they are both *Nymphalidae*). What is true for the long-distance migrant, the painted lady, which in most years only crosses the Alps to central Europe in very small numbers if at all, is also true for the far less conspicuous small tortoiseshell, and for the peacock, which also migrates to a considerable extent in the spring. Their numbers can fluctuate very widely according to the spring weather conditions, with the result that, in the summer and autumn of some years, we see many peacocks and small tortoiseshells, while in others we see only a few. This does not allow us to deduce what we might call the 'native population'. If we want to establish the population of butterflies near us, it is better to exclude migratory butterflies such as the painted lady and the red admiral (especially conspicuous in the garden in summer) altogether and consider the small tortoiseshell and peacock in relation to their spring arrival. Counting butterflies is delightful and eminently important, given the current state of nature, but in order to achieve meaningful results we need to be sufficiently familiar with the lifestyles of the relevant butterflies and properly take these into account when analysing our data (see Photo 17).

To this end I would like to offer a further example from the butterfly world, which, in contrast to the painted lady or the small tortoiseshell, can come into conflict with our interests in the garden and in arable crops, namely the cabbage white.

Poisonous Butterflies and Moths: From the Cabbage White to the Six-spot Burnet

First, an important clarification: there is no such thing as a 'cabbage white', but rather three different species that occur frequently throughout Europe – that is, the large white *Pieris brassicae*, the slightly smaller small white *Pieris rapae* and the green-veined white *Pieris napi*. Other species of white, such as the southern small white *Pieris mannii*, which is very similar to the small white and is becoming increasingly widespread, are for the consideration of specialists and will not concern us here. The three main species provide enough material to discuss the whites in greater detail. We will begin with a rough description. The Latin name of the large white is most appropriate, since cabbage is the principal host plant for its caterpillars, or rather was, until it became comparatively rare. For the large white, alternatives that are used more frequently include nasturtiums, *Tropaeolum majus*, and alyssum species, *Alyssum sp.* The large white is actually considered to be a specialist, feeding on cruciferous plants (*Brassicaceae*), just like the other two whites (see Photo 18).

Cruciferous plants are a diverse and species-rich plant family. It is therefore no wonder that there are considerable differences in their usage by whites. For example, in Bavaria the small white is seen far less often on brassicas in the garden or in the fields; it prefers wild plant species from the brassica family. The list of species that have been confirmed as a larval foodplant for the small white in Germany is more than twice as extensive as that for the large white, and while the large white is principally seen in the open, for example in gardens and around buildings, in south-eastern Bavaria the small white is more attracted to woods and to the edges of footpaths and forestry tracks. However, it is the green-veined white that is to be found almost exclusively there. When it sits on flowers or leaves with its wings closed, the 'green veins' are easy to recognize and a useful characteristic for species identification. The veins of the hindwings appear to have a dusting of green, but are actually covered by very fine yellow and black scales, which is what makes them so striking. The three species form an 'arc of whites' that extends from the garden and the fields over scrubland and the woodland edges all the way to sparse forests. Common to all three is the fact that their caterpillars feed either mainly or exclusively on plant species from the brassica family. These contain toxic glucosinolates, which are passed from the caterpillar stage via the pupae to the butterflies without harming them (as they can detoxify them), ultimately ensuring that these butterflies are distasteful enough to be avoided by birds and are thus protected. The whites can therefore afford to fly around freely despite their conspicuous colouring. They need neither particular speed nor particular care, since once the birds have tasted them and found out how bitter they are, they have learnt a lesson for life. We too value the crops cultivated from the brassica family, such as white and red cabbage or radishes, because of their glucosinolate content. They 'cleanse' our insides of undesired bacteria and are therefore rightly considered healthy.

This incidental remark hides an important point. Why should birds avoid the slow-flying cabbage whites given that the glucosinolates, which they contain, are healthy? This question presents itself every time we consider butterflies that are protected from predation by birds or insect-eating mammals by the toxic substances in their bodies. The answer is that it all depends on the quantity. Other examples include the striking red- and black-flecked six-spot burnet, a day-flying moth from the *Zygaenidae* family (see Photo 19). The *Arctiidae* is another moth family that is equally striking in its colouring and markings and also combines toxicity or unpleasant taste with slow flight. This phenomenon is particularly widespread among tropical butterflies.

Insects that are protected by these 'warning colours' often behave in a 'careless' – sometimes even 'provocative' – manner, as if to say, 'try me and you will soon see what a mistake you have made!' The effectiveness of this optical warning is so high that numerous other butterflies that are not protected through toxins in their bodies have developed 'mimicry' in their coloration, marking or behaviour. More of this shortly. Let us first consider the cabbage whites as examples of toxic, or at least unpleasant-tasting, butterflies in a little more detail. Even though they do not fly quickly, they are still fast and agile enough to evade our clumsy attempts to catch them, as we try to prevent them from laying their eggs on our cabbage, kohlrabi or nasturtiums.

If there is a pair of nesting spotted flycatchers, *Muscicapa striata*, or black redstarts, *Phoenicurus ochruros*, in your garden, it will be easy to observe that these insect-eating birds show no interest at all in the cabbage whites, not even when they are under pressure to catch insects because there are young in their nest crying out for food. Occasionally, a starling will snatch one up, but usually only to drop it again – and afterwards wipe its beak with a distinct gesture of disgust. The protection offered by the glucosinolates evidently works, even for the caterpillars of the large whites, since neither blackbird nor starling, nor the great tit that searches on the ground for caterpillars, is interested in the 'cabbage worms', as they used to be called in south German dialect, when they were still generally well known and, if seen, removed by hand.

Without any natural predators, we would expect the cabbage whites to be exceedingly abundant and to cause enormous damage to crops. We do indeed see them very often, and they are therefore much more familiar than the colourful peacock butterflies. But apart from the local annoyance that they cause in the garden, they are surprisingly inconspicuous. We do not like them, but we do not particularly discourage them, except in those few regions where there are large cabbage fields. I still remember those cabbage fields from my childhood and early youth. As I mentioned earlier, every farming family in the village had some, which they rented out to non-farmers in exchange for assistance in weeding and harvest. One's own cabbages formed part of the staple diet. Anyone who did not have access to a cabbage field would, wherever possible, grow cabbage in their own garden. In those days, in the

1950s and early 1960s, this was still quite normal in the Lower Bavarian Inn Valley.

For a long time now, people have bought cabbage – or even ready-made *sauerkraut* and tinned red cabbage – at the supermarket. Since few people grow cabbage in their gardens these days, cabbage whites are correspondingly less numerous: or rather, the large whites are, to be precise, since the small whites occur everywhere, just as numerous as before. And yet, we have no statistical comparisons with earlier times. Who would have thought of counting cabbage whites at all 50 years ago, let alone in accordance with reliable criteria, for example along fixed transects, together with a record of the availability of cabbage fields? Scant attention is paid to them even now. They seem to be indestructible, virtually unaffected by the mass use of toxins on the fields, over which they still flutter nonchalantly, even though in smaller numbers than before.

Cabbage whites on the Dalmatian coast

I had actually counted cabbage whites repeatedly until the 1980s, albeit for a completely different reason. At the time, it struck me that quite large numbers migrated in mid- and late summer, flying south at the same time as the peacocks and those well-known migrants, the red admirals and the painted ladies. The whites were easy to observe from the large bays of reservoirs along the Lower Inn River, whose banks provided excellent vantage points; by choosing easily demarcated stretches, I could observe which butterflies flew in what quantity across the water over a particular period of time. Either 5-minute or 15-minute intervals were suitable periods for the count, depending on the number of butterflies. Furthermore, there were so many interesting water birds in the area that counting the butterflies was never boring. The results surprised me in two respects. First, the large white made quite regular flights in a southerly direction; second, at the time, I found next to nothing about this in the specialist literature on migratory butterflies. In the handbook *Die Schmetterlinge Baden-Württembergs* it is noted (in 1991, that is, much later) that the large white has a 'pronounced migratory instinct', but that 'according to G. Reich . . . a mass migration was observed from East to West, lasting 5 days, in August 1920'.

The migratory behaviour of the large white was linked to their original natural origins, and this was surely quite correct. They are believed to come from the 'spray zones of the seas, where wild cabbage, sea kale and European sea rocket' grow, which are of a 'truly ephemeral nature' – as it says in the reference book cited above. Under such ecological conditions, the large white would have had to be able to locate and use new habitats quickly and efficiently when it emerged near the beach. This presupposes a readiness to migrate. With the cultivation of cabbage, an incomparably larger feeding area was created in the interior, although, like that at the coast, it was subject to strong fluctuations, since cabbage is not planted in the same fields every year. But this only explains a small part of what happens with the large white. Migrations to new habitats are referred to as 'inland migrations' in the specialist literature. Species that behave in this manner are differentiated from truly migratory butterflies by use of the term 'inland migrants', which refers to those that undertake regional migrations.

However, the result of my counts pointed in another direction. In 1975 and 1979 they confirmed that large whites also undertook long-distance migrations. In August 1979, thousands flew south over the kilometre-wide water surfaces of the reservoirs of the Lower River Inn in the typical formation of migratory butterflies. A few days later, I was walking near the Krimml waterfalls in Austria, now part of the Hohe Tauern National Park. As I was ascending alongside the 380-metre-high waterfall, I was overtaken by thousands of large whites. If a cloud passed over the mountain, as happened often, the whites would rest next to the cascades. As soon as the sun shone on them again, they would fly on. If the shadow lasted a few minutes, I fancied that I would be able to overtake the butterflies, whose easy flight I envied when the footpath by the falls became particularly steep. But then they would swing past me with such agility that I felt more water ran down from my brow than foamed down the waterfalls. When I got to the top, I saw that they hastened on southwards, straight towards the sun. They moved away like a drift of giant snowflakes, supported by the mountain wind, which on that particular day blew towards the south, unlike the *föhn*. Where were they going to, these large white butterflies? Further south to the Dalmatian coast, as in October 1975? That year, I observed a massive migration of cabbage whites on a drive back from Lake Skadar (Skadarsko Jezero) in Montenegro.

I had driven there with friends in order to record the water bird assemblies at Lake Skadar at the time of the autumn migration. This was not the first time I had been there. At that time, there was still an impenetrable border separating Yugoslavia from the extremely communist Albania. It ran straight across what is perhaps the most beautiful lake in Europe. Lake Skadar owed its unique seclusion to the position of that border: it was a dream destination for nature lovers. I do not dare to imagine or verify personally what has happened to that place in the meantime. Perhaps, as I hope, they have managed to preserve enough of the lake, with its north-western sections that reach fjord-like into the Montenegro mountains, full of water-lilies.

The trip to Lake Skadar, from which we were returning in autumn 1975, was not our first. We had already been there in early summer, in the breeding season, and admired the pelicans that floated past the 'Black Mountains' that gave the country its name and out over the water like some kind of prehistoric pterosaurs. However, the October trip was so full of fascinating insights into the nature of the Balkans that we had not anticipated a stopover along the Dalmatian coast road, but were simply driving northwards home. North of Sibenik we came across them, the white butterflies. They came towards us like a loose but distinct cloud. Thousands upon thousands of them. Despite our slow speed the butterflies were crushed against the windscreen whenever the road curved, and this made it easy to identify them: they were large whites. The car's milometer and our wristwatches enabled us to divide up the road into sections for the count that we started straight away. In retrospect it was probably not entirely risk-free, since to the right the cliffs rose straight up to the coastal mountains and to the left the road fell away to the Adriatic Sea without any reliable crash barriers.

The results of the count made the risk worthwhile: when we recorded their distribution over time, the numbers that we plotted for the migratory flight of the large whites looked like the outline of a whale, with a steeply rising, massive head and a body that got smaller and smaller towards the end. This distribution pattern is known as 'Poisson distribution' in statistics. It appeared to be the diurnal mirror image of the results of the transit flight of the painted ladies at the beginning of June 2013, with its 'thick end' towards the evening. However, the difference was more profound. While the painted ladies always travelled alone and clearly did not care whether other butterflies of their species were travelling with them or nearby, the large whites migrate in swarms. We consequently came across them quite suddenly and, within a few minutes, reached the central mass, which thinned out again quite slowly over a distance of several kilometres. Now I understood why the flight over the water on the Lower River Inn was always preceded by something that I had formerly attributed to the need to 'refuel' before embarking on a longer flight. Hundreds upon hundreds would assemble on the (flowering) clover fields of the Inn Valley, near the reservoirs. Up to a dozen cabbage whites could be seen per square metre. Around lunchtime they would start their flight over the water, usually in larger groups: the butterflies did not fly individually like the painted ladies or the peacocks. Accordingly, cabbage whites need much more time to cover greater distances. They probably stop often to rest. I think it is therefore possible that the cabbage whites that I encountered on the Dalmatian Coast were the very same butterflies that had flown south over the River Inn weeks earlier.

Whenever it rains in the desert . . .

Why on earth did the large (cabbage) whites behave like this? Why do they not move south every year instead of just now and then? At the time that I grappled with this question we did not yet know why painted ladies arrived in central and northern Europe in their millions in some years, but then not at all for decades or only rarely. It was only at the beginning of the twenty-first century that the background to this behaviour became clear, thanks to continuous data from weather satellites and the possibility of analysing large amounts of data with efficient computers. It lay in the extent of, or shortfall in, winter rain in the Sahel region. This determines whether the winter generation(s) of painted ladies develop(s) in large or small numbers and thus whether or not they embark on a mass flight northward.

It is highly likely that nobody would have been interested in the painted ladies, which cause no damage to crops in Africa, if they were not closely connected to the migratory locusts, which do. In the Sahel region, the locusts respond in the same way as the painted ladies to the favourable or unfavourable winter rains, and embark on migrations accordingly. In 2003, locusts covered wide areas of northwest Africa. The trade winds gathered them and cast vast quantities of them out over the sea and beyond to the Canary Islands, where they were caught up in a weather system that sent them back in a wide arc to southwest Europe towards the end of the year. They arrived in southern Portugal and Andalusia at a wholly inappropriate time, just as winter was starting, with the result that no crop damage was caused and their appearance before Christmas only led to minor reports in the local newspapers.

Now it was clear, at least, why there had been no mass migrations of painted ladies to Europe for so long, almost for the whole second half of the twentieth century. This was the time of the devastating Sahelian droughts, which cost the lives of many people and vast quantities of their livestock, while bringing us cold, wet summers in central Europe, since the rain fronts in such years travelled almost parallel to the Alps towards the northeast. This damp weather benefited the large white. The small white too, but for them the changes were not as conspicuous as for the large white, as shall be explained shortly.

Which factors affect the reproduction of butterflies, and when?

Naturally, the weather alone was not, and is not, responsible. Temperature and precipitation encourage plants to thrive, and although they sometimes damage them, the consequences are not discernible to us, provided that they remain within the normal range for the local climate. In the case of the painted lady, the determining factor was probably the fact that the abundant winter rain allowed the vegetation to flourish across a very wide region that had been overgrazed and affected by persistent drought. The intensive pesticide campaign against the locusts that had taken place recently probably devastated a significant portion of them, since DDT was sprayed extensively from aircraft. With our large whites, the circumstances surrounding the population explosions are not nearly so clear. Weather can certainly not be held responsible for the fact that they subsequently became rarer (convenient as this might be).

The wet summer of 1975 may have contributed, but ought not to have been the decisive factor. In the 15 years prior to this, and into the early 1980s, I regularly observed very large gatherings of cabbage whites and followed their flights over the reservoirs. More incidentally than deliberately, but nonetheless with a certain astonishment at what I saw. In the 1960s, I actually became attracted to cabbage whites with a certain youthful curiosity, since their caterpillars were eating the cabbage and the kohlrabi in our garden. The damage that they caused was evident but, at the same time, did not prevent the development of normal heads of cabbage or kohlrabi stems. When the yellow-spotted and not particularly beautiful caterpillars were ready to pupate, many of them crawled up the side of the house, as I briefly described earlier. However, strangely enough, a significant proportion remained on the path, which is to say, they did not manage to reach the place under the shutters where the others were. Those under the shutters pupated. Those that had remained on the path did not: on the contrary, something very remarkable happened to them. They remained in the place that they had reached, stretched out, but did not move any further. A couple of days later, they would be covered with a dozen or more small yellow cocoons that seemed to be stuck to the caterpillar. The larvae that had pupated inside them were from the Braconid wasp, *Cotesia glomerata* (formerly *Apanteles glomeratus*), that I mentioned earlier. The small larvae of this inconspicuous black mini wasp had developed within the bodies of the caterpillars and gradually consumed them, while the caterpillars continued to grow. By the time the caterpillars were unable to crawl any further, the parasites must have eaten into their muscles. Now ready to pupate themselves, the wasp larvae bored their way out through the caterpillar's skin and made themselves long egg-shaped cocoons.

After some time, depending on the temperatures that occurred while they were developing into small wasps within the cocoons, the Braconid wasps emerged and set out to look for young caterpillars of the new large white generation that were starting to grow. Two or three such cycles can take place within one generation of caterpillars. We called them 'caterpillar' eggs (in those days) as we found them practically all over the garden and saw them on the walls of buildings, mostly slightly apart from the cabbage white caterpillars that they had destroyed. I bred large cabbage whites in large numbers in the

1960s, without great enthusiasm, in order to discover how great the extent of the parasitization was. I lacked enthusiasm because I did not want to simply kill the butterflies that had successfully emerged. They were without doubt beautiful when seen up close, with the undersides of their hindwings a creamy yellow. However, propagating cabbage whites was also unthinkable. Farming neighbours, in particular, would have had little sympathy for such an eccentric activity. Cabbage whites were pests, and that was that!

I do not remember exactly how I extricated myself from the situation, but in any event I obtained clear results: the rate of parasitization was high, very high, except in years when there were mass gatherings of large whites on the clover fields along the Inn Valley, from where they flew away over the water. Somehow or other they must have avoided the parasites to such an extent in those summers that the losses they normally suffered in the first generation turned out to be small. The second generation was thus able to achieve a mass reproduction. Whereas, for the painted lady, it was the weather and the quantity of rain, for the large white it was the loss in population growth through parasitization that was the main factor governing population control. In other words, where for the former it was a factor caused by the inanimate environment – that is, an abiotic factor – for the cabbage white it was a biotic one. This finding did not simplify our interpretation; quite the reverse. On the whole, it is the abiotic parameters that are easier to understand and (for the purposes of computer analyses) to model than the biotic ones. This is because everything inanimate has a direct and general effect and does not depend on whether the species affected is common or rare. In ecology, this is known as 'density-independence'. The effects of one group of living things on another will, however, depend on how numerous each group is in relation to the other (at the time). These effects are therefore known as 'density-dependent'.

The difference is extremely important because density-dependent factors have a controlling effect. They curb strong population fluctuations that would otherwise be inevitable, since the increase of one species follows the corresponding growth of the other, albeit at a different speed. In contrast, if painted ladies or migratory locusts become more numerous due to heavy rain in a particular winter, the weather in the following year will not become drier in order to 'balance things out'. To put it another way, weather and climate are not concerned about their consequences.

Useful models

Experiencing such differences at first hand is extremely important for ecologists. However, as convincing as something may be as a model, it still needs to be proved in real life. There are good reasons why one says that 'all theory is grey'.* Too much trust should not be placed in theoretical concepts or models that are declared 'laws'. Nature, especially living nature, seldom observes such guidelines. They often show themselves to be little more than wishful thinking and shatter when confronted with the complexity of nature, which has been over-simplified for the creation of the model.

I had to learn a wide range of ecological laws, rules and concepts in the course of my studies. Some of it was difficult to understand since it was expressed in our textbooks in terms that were too academic. A sentence from a large ecology textbook remains in my memory, and I have often cited it: 'The situationally variable realization of the strength of vagility produces mobility.' This was textbook-style language, which had to be thoroughly digested before its contents became fit for purpose. It merely meant that the animals in question rambled around either more or less, depending on the situation. North American and British textbooks offered better alternatives to this tough diet, but they were not necessarily any livelier. They gave a similar weight to models, particularly mathematical models, as the German textbooks gave to such a highly scientific style.

When I was studying at the University of Munich in the late 1960s, we still had immense freedom. In fact, there were hardly any restrictions. In May 1968, shortly after the beginning of the summer term, I was confronted with a population explosion of moths that still make the headlines today: small ermine moths. A closer look at them will reveal how naive it is to assume that research findings that have been confirmed again and again will lead to changes in public perception of

---

* From *Faust, Part One*, by Johann Wolfgang von Goethe (1808): 'All theory, dear friend, is grey, but the golden tree of actual life springs ever green.'

natural phenomena, never mind influencing the measures that are taken to counteract them in any meaningful way. In the area of so-called 'pest control', age-old prejudices still prevail.

The need to go slow

However, first it would be worthwhile to take up the issues that have been raised and to extend our understanding a little. One of the very big differences between painted ladies and cabbage whites is that the latter are protected (from birds) through poisonous, bad-tasting substances – but not the former. This might lead one to conclude that painted ladies need to fly very fast, otherwise they will too often fall prey to birds. Their speed of flight has to be so fast that they elude even the swiftest birds; the birds, in turn, learn that, since their efforts are only some-times successful, the energy expended is simply not worthwhile. This must be right, since 'nature', people say, proceeds in economic fashion. Only butterfly collectors who do not depend on the successful catch, but pursue it as an 'affordable luxury', will chase after swift butterflies. Even well-respected writers subscribe(d) to this opinion, as I will detail in my final chapter in this book, 'The Beauty of Moths and Butterflies'. Nevertheless, this absurd-sounding observation is reasonable, since the cost–benefit calculation does represent something like the Holy Grail in the significant area of ecology that deals with nutrition and food acquisition. Let us therefore take the concept as our departure point for some concluding considerations on cabbage whites.

They fly slowly, much more slowly than the painted ladies. They therefore need much more time for long migratory flights. This exposes them even more to the inclemency of the weather on their journey. The weather does not consider whether the butterflies are common or rare at any given time. In short: the likely losses suffered during a migration will affect whether a migration is undertaken or whether it is limited to particularly favourable conditions. Incidentally, this applies equally to birds and bird migration. We shall now return to the issue of flight speed and ask the question again, in a slightly different way: why do the large whites not simply fly as fast as the painted ladies, which they resemble so closely in body size and shape that even butterfly experts could probably not differentiate between them without their colour-giving scales? The large whites ought to have just as many chest muscles for the flight and just as much body fat for fuel as the painted ladies. To ask such a question is not absurd. In fact, on closer considera-tion it contains two interesting aspects.

First, take the direct performance comparison of the two large but-terflies. One of them, the painted lady, flies fast and far, while the other does not. Second, consider the question of a (worthwhile!) goal. The painted lady clearly has a fixed goal, since it makes the migration even if its population is not very large and the flights only take it as far north as the Mediterranean area. This produces a smaller migration, but still allows the butterfly to avoid the desert climate with its unbearable summer heat. Even if normal summer temperatures are reached around the Mediterranean, this heat is significantly lower than the tempera-tures south of the Sahara, and there are plenty of summer storms with showers around the Mediterranean too. The migration over the Alps is a kind of logical extension of a small to a large migration, when there has been a population explosion south of the Sahara. Here the goal is clear, and by no means just for the painted lady that we are using as our example, but also for the red admiral that migrates much more regu-larly between the Mediterranean region and the regions north of the Alps, or the small, day-flying hummingbird hawk-moth, *Macroglossum stellatarum*, which is mistaken for a hummingbird every now and then. The same applies to a whole series of other, less well-known migratory Lepidoptera such as the small, narrow-winged *Nomophila noctuella*, or its huge, beautiful counterpart, the tropically coloured oleander hawk-moth, *Daphnis nerii*. I could add many other insects that undertake similar migrations between the Mediterranean region or North Africa and central to northern Europe, in quantities that differ widely from year to year, and even some fly species. In favourable summers, the air is full of such flyers. Common to all is the fact that they have a goal that, from our perspective, is rational or at least plausible. Vegetation north of the Alps, in a moderate climate, that has been watered by heavy summer rain offers better conditions for reproduction than land that is characterized by heat and summer dryness around the Mediterranean. The Mediterranean climate is famous for good reason. The other factor that is common to them all is that they suffer heavy losses during their migrations and these are only 'worthwhile' if favourable air currents contribute to an increased flying speed and a shortened journey time. There are therefore good and bad summers for the migrants from the south.

However, with the large white it is difficult to attribute a goal. Where would it want to fly to? How could there be any profit for it in the relentless struggle between outlay and return? The observations set out in the reference book for the butterflies of Baden-Württemberg cited above even appears to contradict my own findings. It says that the mass flight in August 1920 took place from east to west, not towards the south. But perhaps the answer lies in this apparent contradiction. If those large whites that flew west for five days in August 1920, just like those that crossed the River Inn in southeast Bavaria in August in later years, travelled further, they would quite reasonably, within about a week, have reached the Atlantic coast. Those travelling south from southeast Bavaria would have reached the shores of the Mediterranean or perhaps even the north African coast. Both would be 'prudent' goals for a species originally assumed to come from coastal areas near the sea, where the wild forms of cabbage and radish grow. However, despite good protection from birds, the migration involved heavy losses and would therefore only have been worthwhile if those large whites that remained behind fared even worse than those that migrated. Their pupae would hibernate in the temperate climate of central Europe, which would have been out of the question for painted ladies in their butterfly, caterpillar or pupa stages, and only exceptionally successful for other migratory butterflies such as the red admiral. It is easy to imagine that after a successful reproduction, such as that which took place before the mass migration, the pressure from parasites became so enormous that those which remained in place hardly had any chance of survival.

The question that remains open here may be expanded by considera-tion of the small ermine moth. The note from the Baden-Württemberg handbook on the former natural occurrence of the large whites in coastal areas is of direct relevance: 'The spray zones near oceans with wild cabbage, sea kale and European sea rocket were of a very transient nature. A species that tries to live here must be an r-strategist, that is, one which demonstrates a quick transition from one generation to the next with a high number of descendants, in order to seize newly emerged habitats quickly but also to be able to rapidly make up for large losses through seawater, etc.' Mass reproduction promotes migration. Habitats that are only available for short periods can only be colonized by those species that are sufficiently mobile. This is the message of the above formulation. Being an r-strategist means that the species invests in as many descendants as possible when it reproduces so that the local population grows quickly and may rapidly exhaust the food supply. This leads to wide population fluctuations. Their counterpart, the k-strategist, adapts its propagation rate much more closely to the available resources, with the result that it can survive in one place for longer and experience smaller population fluctuations.

Hence, the migratory flights of the large white may not have been regular migrations corresponding to those of the 'true migratory butterfly', but rather so-called 'dispersals', that is, movements away from a reproduction centre where there is a large surplus. This can lead to resettlement. The general direction of flight may be oriented towards the coast, since that is where suitable habitats naturally occur. We may therefore consider one part of our question to have been answered. The conclusions produced can be objectively checked. The second part of the question deserves separate consideration because it affects many Lepidoptera and other insect species in more or less the same manner. This is the question concerning the connection between toxicity and slow flight.

Poison in the body

Let us turn the question around and rephrase it as follows: how can certain butterflies and other insects store toxins in their bodies as protection against being eaten without poisoning themselves? Here, surely, is where the larger problem lies, the one concerning mobility. If it was simple to transfer protective toxins from the caterpillar via the pupa to the butterfly, then many more species – or practically all of them – would be poisonous and protected. Since this is clearly not the case, and since we humans are also obliged to pay attention to which plant foods suit us and which do not, those that are resistant to toxins must have something that puts them in a position to live with poison in their bodies.

Two different ways of dealing with toxic substances in food are plausible and we can find examples of both in butterflies: either the toxin is chemically converted by means of special enzymes into a different state that is no longer toxic, but can promptly be returned to the toxic state, or it is transported and stored so quickly that it does not harm the caterpillar. The vast majority of insects that live off poisonous plants, and almost all plants are poisonous in some way or another, rid themselves of the toxin through enzymatic conversion and chemical decomposition. This not only makes the toxin ineffective but enables its excretion. They have thus overcome the difficulty, but have not gained anything from it, since without the toxin in their body they are no longer protected.

At this point it would be tempting to go into the chemistry in detail, since it is closely connected to the formation of pigments and many Lepidoptera and their caterpillars are conspicuous because of particular colours. However, that would go beyond the scope of this book. We must be content with the clarification that most toxins contain compounds that are chemically complex and that the toxic effect is often connected to sulphur and cyanide compounds. With the cabbage whites, the compounds in question are the sulphurous glucosinolates that I have already referred to, while with *Zygaenidae* moths, such as the six-spotted burnet, other substances release highly toxic hydrocyanic acid. However, they are subject to the general (bio)chemical rule that the effect of chemical substances and the speed of chemical reactions are dependent on temperature. As humans, we are relatively sensitive to poisons for this very reason, since our metabolism functions intensively at (almost) 37°C. However, birds, even small songbirds, surpass us significantly in this respect. Their internal body temperature can be as high as 42°C, which means they live just below the fatal limit. Following rapid flight, and if external temperatures are high, the body temperature of insects can also reach 40–42°C for short periods. Their metabolism will then run at full tilt. Now, since the effect of toxins increases with rising temperatures, butterflies that carry toxins must be restrained in their movements in order to avoid generating an excess of heat. Their habitual slowness saves them from autointoxication and prevents them from flying faster or for longer periods. Since all poisonous insects behave in this manner, there is much to be said for this interpretation.

It even applies to poisonous snakes, which are sluggish, relatively slow and only fast for short periods if they need to be. Poisonous toads also crawl more slowly than nontoxic frogs. The general implication is that poison slows you down.

With poisonous species, all other activities that relate to a heightened conversion of energy are thus also diminished or suppressed. The (major) advantage of protection from predators brings limitations in other areas. Accumulating toxins in the body is by no means the best option, although being faster than the pursuer also entails huge (energy) costs that could be spent in other ways (for example, in egg production). Parasites manage to circumvent the toxin defence anyway, for example by not eating tissues such as fat cells in which toxins are stored in the caterpillar's body. Being conspicuous by means of warning colours, or by eating plants in exposed places, can therefore place the butterfly at a disadvantage if the parasites are able to recognize such colour signals and feeding patterns. In the interplay between predator and prey, neither ultimately wins the decisive advantage. From our theoretical perspective on nature, this situation must be prevented in any event, since the overefficient parasite would either eradicate its prey and therefore the basis for its existence, or the prey, released from predation pressure, would use up its food too quickly and thus be responsible for its own demise. We do not know whether nature adheres to our ideas or not. Extinction does take place, always and everywhere, but mostly goes unnoticed. Nonetheless, the fact that the (agro-)chemical fight against pests seems to aspire to the total destruction of the pest species that it targets should give us food for thought. The speed at which the opposite is achieved, that is, resistance by pests to pesticides and new mass proliferations of the pest species in question, has led to an abstruse race: between the pests, which are constantly getting faster and better, and the chemical insecticides, which are constantly being redesigned. This problem is aggravated by the extreme genetic standardization of crop plants, since it has long been known that mass proliferation and the development of resistance in pests is best controlled by genetic diversity in the relevant crop. If the situation were not so serious, one could simply shake one's head in despair.

The problem of the overabundance of large whites and their damage to cabbage cultivation or the garden was largely solved by itself, or in a more natural manner than through the application of pesticides. Over time, less cabbage was cultivated, so the large white lost the extensive food supply that had previously been available. Add to this the virtual disappearance of the clover fields, which the large whites had been using as 'mass refuelling stations' on the longer migrations in mid- and late summer, when farmers stopped fertilizing their fields with clover and started to apply manure and slurry. The extensive loss of the clover fields had an adverse effect, not only on the large whites but also on the clouded yellow, *Colias croceus*, and others that used to be characteristic of meadows and fields. A reduction in a food source generally has the greatest effect on the frequency of occurrence and population development of the species that live on it. Thus, the fly population dropped sharply when water and manure removal was introduced in cattle sheds. Fly larvae that had formerly flourished inside dung heaps could not survive in liquid manure. The installation of compost bins in gardens has led to an increase in certain species of fly, but by no means to the same extent as used to be generated by manure heaps in cattle farms. If nettle plants were eradicated on a large scale, then nettle-loving butterfly and moth species would become rare. These are indeed two examples of situations where available food and changes to that availability determine abundance and occurrence of species, although in nature this approach towards control can only rarely be implemented. Usually it does not work, since the crops concerned need to be maintained and kept productive. It is much more common for the control of injurious insects to be regulated by the next most efficient factor, usually natural predation, above all by parasitoids. How they work is seen most strikingly with bird-cherry ermine moths. If they have infested bird-cherry trees, we can simply relax and watch the process, since no lasting damage is caused and we can learn much from it.

The Secret Life of Small Ermine Moths

Trees on flood plains or in riverside forests that have been stripped bare of their leaves and are covered in silvery-white, web-like sheets in the middle of May, when all around is turning green, are a strangely beautiful sight. These trees, which have sometimes been eaten down to their last little leaf, seem so lost and helpless that they set off our internal alarm bells. And since the webs extend down to the ground vegetation, the whole wood looks as though it has been devastated. However, what in the May sunshine seems somehow striking, or indeed disarming, against the bright blue sky, swiftly loses its sinister charm after the prolonged heavy rain typical of this month. The webs, silver and shining only in fine weather, then become discoloured, turning a dirty grey-brown colour when the caterpillar excrement trapped within dissolves, rendering the denuded trees even more of an eyesore. This reaction is understandable, but not necessary. In a matter of weeks, usually on the threshold from May to June, the trees put forth fresh green leaves. Later, in high summer, only the remnants of the webs on the trunks provide any clue that the trees had been stripped bare in May. The new leaves actually look even better than those that were devoured in the springtime: most of them are perfect. They will become brown in autumn as usual; dry out and fall to the ground. The complete defoliation of the trees, which briefly resemble skeletons, is followed by an almost ideal condition, as if nothing had happened to them. How can this be? Why do we only see this phenomenon in some spring seasons and not every year? As we will shortly see, these simple-sounding questions are not at all easy to answer, and certainly not through investigations that are only carried out over a few years. I have been researching ermine moths since May 1968, that is, for almost half a century to date, and there are still important questions that remain unanswered. What I have learnt about the ermine moths in these 50 summers can be summed up in two key statements. First: many years of investigation are insufficient for understanding the population dynamics of moths or butterflies; often, even a decade is too short. Second: the interactions between insect and plant are much more profound and complex than we like to imagine, particularly in the area of pest control.

The bird-cherry, a tree of the riparian woods

Below I shall attempt to explain the most significant findings. Almost everything can be followed directly by anyone without any great technical effort. Let us start with the observations briefly delineated above. We see trees that have been covered in webs and more or less stripped bare. Invariably, these trees are bird-cherries, *Prunus padus* (see Photo 20). This is the most important finding, which immediately soothes our fears: only this tree is affected, and none of the other tree species in the riparian woods. There may be silk-spun bushes of spindle or hawthorn, however, in the hedgerow or in the open. These quite different web coverings are caused by other moths. The bird-cherry is a tree species typical of moist flood plains. It grows in stream valleys and riverside woods. In years without heavy ermine moth infestations we can see and smell them when they blossom in late April or early May. Their creamy white flower clusters formed of many small rose-like florets are conspicuously displayed against the young May green. In the occasional years in which they flower with particular abundance, the bird-cherry trees appear as though covered in foam.

The unmistakeable fragrance carries well, although its distinctiveness sadly eludes description, since there is no widely understood classification of smells. Anyone who appreciates wine will know this. The bouquet of a particular wine, as described on its label or in advertising, is often closer to creative fiction than a reality verifiable by sense. This is not intended as criticism: I am merely drawing attention to the shortcomings of our methods of differentiating and classifying smells. Quite apart, that is, from those stenches that dogs, rolling around in the source of the questionable smell, will attempt to transfer to their fur, finding them just delightful. This digression has a valid point that can be confirmed by anyone who has ever attempted to take home the fragrance of those bird-cherry blossoms that smell so special in the forest. The flowering sprays bear up well in a vase with enough water, but after a few hours, and certainly by the following day, the fragrance that was so enchanting outdoors will have changed into a truly unpleasant odour that can cause headaches in sensitive people.

Whatever it was that characterized the atmosphere of the spring forest, we experience it quite differently after it has been in an enclosed room for a time. However, the smell remains unmistakeable. If we wonder where it comes from, the superficial answer is 'from the blossoms', but as true as this statement is, it tells us little. The flowers do not simply produce fragrance: they can only emit what the tree, or the particular plant, has provided as raw material. This consideration brings us to the core point. The bird-cherry is a species of cherry tree and, as such, is closely related to our sweet cherries and the wild, sweet or gean cherry, *Prunus avium*. Just like sloes, these belong to the genus *Prunus* = plum. The bird-cherry trees do develop actual cherries, but these remain small in comparison with wild cherries and appear black and shiny when ripe. They are edible, but, as well as being sweet, they contain an astringent, bitter substance. Little flesh surrounds the stone and they are therefore not highly valued as wild fruits; rather, they are frequently assumed to be poisonous. Sloes, which are also a kind of plum, develop higher levels of bitter substances and seem more poisonous, so that we cannot enjoy them unprocessed as we can the wild and cultivated sweet or sour cherries.

Toxins in bird-cherries

This consideration of the cherry does serve a purpose. What we barely notice about the fruit of the bird-cherry is in fact of vital importance to the ermine moth. It is clear from the blossom fragrance of the bird-cherries that we are dealing with ingredients that lend a special, entirely unmistakeable smell to the leaves and the thin twigs. This smell is released when we crush a couple of leaves between our fingers. For most people, the smell is unpleasant or repellent. In any event, it does not inspire one to steep the leaves to make tea, which, given their toxicity, is just as well. The leaves contain amygdalin and iso-amygdalin, which are highly toxic hydrogen cyanide glycosides. The toxic effect comes about when the hydrogen cyanide element is separated.

I have already referred to this in the section 'Poison in the body' in the previous chapter. These hydrogen cyanide glycosides are also present in bitter almonds and other seeds of stone fruit as well as in apple seeds, all of which are therefore poisonous. The hydrogen cyanide is removed in the stomach and the intestines. And this brings us back to the ermine moths, whose caterpillars live exclusively on the leaves of the bird-cherry. They are therefore quite correctly called bird-cherry ermine moths, although their scientific name, *Yponomeuta evonymella*, is unfortunately wrong and misleading. Wrong, because the species name *evonymella* refers to the spindle-tree, *Euonymus*, on which the bird-cherry ermine moths do not live. It is caterpillars of another ermine moth, the spindle ermine, *Yponomeuta cagnagella*, which they resemble as moths, that live on this tree. We will revisit their characteristics in connection with the ermines later on.

We must bear in mind that the bird-cherry ermine moth is to be found only on the bird-cherry tree, and not on any other species of tree or bush. The confusion with the scientific name is due to reasons that still challenge us today. Ermine moths are small, long, narrow and silvery-white. On their forewings they have longitudinal rows of fine black dots. The moth that lives on the bird-cherry has five such rows; the one that lives on the spindle-tree and the hawthorn bush has three. It is almost impossible to differentiate between the species that occur in central Europe if one is not intimately familiar with the differences between them, such as the fine rows of dots, the colour of the fringes at the edge of the wing and whether these carry a dark stripe or not, and if one does not also happen to have a magnifying glass to hand for closer inspection. As a group, they represent an ecological object lesson in the diversification of a genus across a spectrum of similar yet distinct species with narrow ecological 'niches'. Nature has many such impressive examples, allowing us to study evolution and ecology. Textbooks do not always have to copy the same standard examples that we see in the Anglo-American specialist literature, and which may exist only in distant regions.

In our case, we are concerned with the specific ingredients of the bird-cherry, which can be tolerated as a foodstuff only by very few insects. If we inspect this tree a little closer, we will generally find barely more than a handful of different insect species. Among them will be the caterpillars of the clouded magpie, *Abraxas sylvata*, from the family Geometridae (German name '*Ulmen-Harlekin*' or 'elm harlequin'). They, however, tend to prefer elms, as the German name suggests, over bird-cherries. This moth, characterized by round rust-coloured spots on silky pale wings, is found quite rarely on cherry, compared to the bird-cherry ermine moth. More often, we will find aphids and cone-shaped growths on the leaves, from gall mite infestations. It is only the gall mite species, *Eriophyes padi*, that makes its home on the bird-cherry tree. And that is about all. No other species of tree native to the central European woodlands has so few species of insect living on it as does the bird-cherry. With oaks, the number of species goes into the hundreds, even though oaks have a level of tannins that should actually keep herbivores away. In practice, these are generally 'curbed' rather than kept away altogether, so that oaks are seldom bare. If they are, this can affect the tree very badly. Population explosions of the green oak tortrix or green oak moth, *Tortrix viridana*, and the oak processionary moth, *Thaumetopoea processionea*, whose caterpillars are feared due to their toxic and irritating hairs, can really damage an oak. As I write, in spring 2018, there is once again a very bad infestation of the latter in the oak forests of various regions of central Europe. North Bavaria is particularly badly affected. The wind disperses hairs that have broken off the bodies of the caterpillars, and as a result many people suffer from skin irritations and respiratory difficulties. The mass outbreak can be ascribed to the very warm weather of 2015 and the years that followed, although the true causes are evidently unknown and have yet to be adequately researched.

The very small number of species that live on the bird-cherry is sometimes explained by the fact that this tree only arrived late in central and north-western Europe, after the end of the last ice age. It survived the ice age in the refuge offered by southeast Europe and the near East, just like many other species of tree and plant. But while this may be a plausible explanation for the situation in Britain, the final north-western area reached by the bird-cherry, this idea is not convincing when applied to the paucity of insect diversity on this tree species in central Europe. Why would it not have brought the normal spectrum of insect species with it on its return after the ice age? It is not as if the bird-cherry was artificially introduced to central and north-western Europe, like the horse-chestnut, whose 'moth', *Cameraria ohridella*, the horse-chestnut leaf miner, has made the headlines in Germany in the last two decades, since it appeared to pose a threat to Bavarian beer garden culture. It is far more likely that the answer lies in the special toxin, the ingredient that produces the characteristic and, to us, repellent smell.

It is clear that only very few insects can cope with it. This is illustrated by the birds that eat caterpillars. Even the cuckoo, which consumes even the prickliest caterpillars and sheds its whole stomach lining to get rid of the accumulated hairs, disgorging a kind of pellet, does not favour the caterpillars of the bird-cherry ermine moth, although they are often available in thick hordes. The golden oriole, which is not exactly fastidious about tropical insects full of defensive substances, also seldom eats the caterpillars from the bird-cherry tree. Varying reports may be based on observations relating to ermine moth caterpillars that are less poisonous or nontoxic and that may greatly resemble those of the bird-cherry ermine moth.

The caterpillars of the bird-cherry ermine moth can evidently tolerate the toxic ingredients of the leaves of that tree, and, indeed, so well that they sometimes strip the trees bare. Let us have a slightly closer look at their lifecycle: it will turn out to be very instructive.

The life history of the caterpillars of the ermine moth

At the same time that the bird-cherry produces its first leaves in spring, the caterpillars of the ermine moth hatch from the eggshells in which they have spent autumn and winter safe and snug. Small as they are, their jaws are still very weak, and they can only bite and consume very young and tender leaves that have not yet hardened. They do this from within the spring shoots in which they have nestled, and from where they are protected from any inclement weather. There could be a late return of winter weather bringing frost, in which case many of these newly hatched young caterpillars will perish. It is never all of them, since there are always some that do not hatch with the first leaf buds, but weeks later, in late April or sometimes early May. They usually escape late frosts, although not every year, as the 'Ice Saints'* occasionally bring frosty nights as late as the first half of May. It is extremely important to differentiate between 'early' and 'late' caterpillars, since the latter will represent the reserve in the event that the former perish. However, the latter will have no opportunity to survive and pupate if the weather is favourable for the little early caterpillars, who will have a head start in terms of development and can bring about complete defoliation. In this case, little or nothing will be left for the later ones. They will become the 'starving caterpillars' whose activity we will come to shortly. The important thing to understand is that both are essential, since weather conditions vary greatly from one spring to the next. The advantages of favourable April weather can be replaced all too soon with the fatal consequences of unfavourable weather in May.

After their first moult, the caterpillars continue to eat the leaves of the bird-cherry, which have, in the meantime, become larger and more abundant, and will start the work of fabricating many-sided tent-like webs. If the weather turns bad, they retreat inside these, including during the cool May nights. If the infestation is small, it is because not many ermine moths hatched in the preceding summer and mated, and only a few females laid eggs; in this case, we will not see many of these ermine moth caterpillars. We might see a web at the end of some of the twigs in the middle or lower levels of the bird-cherry tree, but it will soon disappear behind the enclosing foliage, which grows much thicker and greener on bird-cherries than on the surrounding alders. If many females have successfully laid clutches of eggs on the bird-cherry buds that have formed, ready to open next year, then the infestation will be visibly larger.

The foliage of the bird-cherry quickly becomes sparse as the webs

---

* St Mamertus, St Pancras and St Servatius, whose feast days fall on 11, 12 and 13 May, respectively, and who, according to folklore, bring the last of the cold weather.

become thicker and more extensive. By the middle to the end of May, the first trees can become covered in webs and are stripped bare. Where they stand together in groups, in places where several trees have developed from the same rootstock, defoliation and webs will affect them all. However, alders, ash trees or other forest trees will never be eaten, as these caterpillars specialize exclusively on bird-cherries. They are monophagous, as the technical jargon puts it, that is, adapted to a single food source, and, as can be seen, they have done this very successfully. In their masses, they crawl from the crowns of the trees that have been stripped bare, down the web-covered tree trunks, to a fork in the branches or the foot of the tree, and search for a relatively protected place that is suitable for pupation. It should, in particular, be protected from rain and hail. Here they will cluster together in their hundreds, even thousands. In a few days, the pale yellow, black-spotted caterpillars will turn into narrow, spindle-shaped barrels. They have woven themselves into these cradle-like cocoons and attached them closely to each other. But then something unique tends to happen. Cadaverous-looking caterpillars cover these accumulations of pupae with a thick outer web. This will protect them from rain and inclement weather. Moreover, it prevents ichneumon wasps and tachinid flies from attempting to lay their eggs on the pupae; the larvae of these parasitic insects would consume them whole. If the exterior protective web is dense enough, then it will be difficult or impossible for parasitoids to crawl through it. In the course of my work at the Bavarian State Collection of Zoology (ZSM), I made many investigations of the pupae masses in the protected centres of such webs, some of which produced phenomenal emergence rates for the moths of up to 98 per cent (see Photo 21).

In years when defoliation occurs, with bird-cherry trees becoming covered in silver-white webs and appearing almost skeletal by May, a mass emergence of small ermine moths will take place in June and July (see Photo 22). In midsummer, one can see them sitting everywhere in the forest, on the vegetation or on tree trunks. If one approaches these little moths, they do not usually fly upwards but instead jump off and fall into the dense ground cover. Logically, such a mass emergence ought to be followed the next year with another population explosion and defoliation. However, this only happens very rarely, and then only when the preceding summer was a 'good' one for the ermine moths, but not an exceptional one. The question of why this might be has occupied me for several years. Before going into it any further, I would like to return to the caterpillars that produce the outer web that protects their fellows within the pupae that they have already made.

Helpful hungry caterpillars*

As mentioned above, these caterpillars do not look good. They are called 'hungry' because their bodies are usually significantly thinner than their head capsules. Their movements seem slower or more 'tired'. According to my investigations, none of them ever managed to make it to pupation, even when I took them home and kept them there. The reason for this is visible in the trees above: the other caterpillars that had pupated successfully had eaten the branches bare so there was too little food left for the 'hungry' caterpillars. This was probably the normal fate of the 'late' caterpillars in spring, if the weather turned out favourably for the 'early' ones. Yet, if they are faced with wintry weather late in the season, the 'late' ones will triumph, since enough food is left for them in the foliage that has probably been barely touched.

As hungry caterpillars without the opportunity to pupate and, more importantly, without the materials stored within their bodies that the females will need to produce eggs, they really have nothing left to lose. But since they are possibly, indeed very probably, closely related to the successful caterpillars, because they may have emerged from a clutch from the same female, their actions improve the survival chances of their siblings. The overwebbing of the pupae mass by the caterpillars that are doomed to die works in their siblings' favour. The hungry caterpillars are not even fit for parasitization by parasitoid wasps. At least that is what I strongly suspect, although I have not yet carried out the appropriate research. I never saw the parasitoid wasps show any interest in the hungry caterpillars. To be sure of this, I should collect the hungry caterpillars and check to see if they are ever attacked by parasites and, if so, by which ones, and how their emergence rates turn

---

* 'Hungerraupen' is used by German entomologists to refer to larvae that do not get enough food for pupation because conspecifics have already eaten everything.

out. As indicated at the start of this section, we are still a long way from understanding everything about the complex relationships between ermine moths and bird-cherries.

Between parasitism and population explosion

The finding that the extent of parasitism increases with the number of ermine moths, but then drops off rapidly with very high populations, is well established. If there are very few, then the degree of parasitism remains low. If the caterpillars are quite numerous or very numerous, then the degree will rise and can exceed 50 or even 70 per cent. But in the case of mass outbreaks with defoliation, it falls away so abruptly that only peripheral pupae are affected, which therefore has only a slight effect on the whole population, if any. Emergence rates of 98 per cent of ermine moths do occur. The adults come flooding out of the pupal masses on summer nights as if effervescing. It raises the question of why the restraining effect or proper regulation by means of the parasitic ichneumon wasps, Braconid wasps or the tachinid flies that look like prickly houseflies fails to function just at the time when the ermine moths have become particularly numerous. The loss rate to natural enemies is actually lower during population explosions than in phases of low population.

One particular discovery made the whole matter particularly unclear. It was shown after a mass outbreak that the moths that had emerged successfully were significantly smaller than those from a normal year with a mid- to low level of infestation in the bird-cherry trees. Closer investigation revealed that it was principally the females that were affected. They had far fewer eggs in their abdomens than when populations were lower, or they had none at all. This shortened their body length by up to 2 millimetres. The mass emergence thus feigned success but was, from a biological and reproductive perspective, nothing of the sort. What was the point, if the moths that developed from a majority of the caterpillars had severely limited fertility or indeed none at all? This appeared to be the explanation for the fact that years in which defoliation took place and the emergence success was high were followed, despite low levels of parasitization, by lower or very low infestation in subsequent years. Observed over the course of decades, groups of years with cumulatively higher occurrence of bird-cherry ermine moths clearly stand out from those with low infestation.

The frequency of population explosions has not generally increased in the past 50 years, despite some suggestions based on 'impressions' that were believed without consulting the available quantitative investigations on the matter. The current increase in the small ermine moths would indeed fit all too well with climate change, although this cannot be used to explain every change. In the late 1960s and early 1970s, there were actually more years with severe ermine moth infestations than in the years since 2010. Yet the question posed in the context of a trend, whatever that may be, only deals with one side of the phenomenon. The other is contained in a more important question: what is the significance of 50 years in respect of the development of the population of a moth species that is so narrowly specialized? How does the extreme increase and decrease really come about? This has still not been satisfactorily explained.

Longer-term population cycles

Let us take a step back and consider the process from year to year. As I have already emphasized, and as we can readily observe outside, there are years with high infestations, those with medium to weak infestations and, finally, those in which ermine moths need to be searched for if they are to be seen at all. Also, as we have just ascertained, the level of parasitism is highest where the population size is medium. We can assume from this that the ermine moths, or, to be more precise, their caterpillars, are more successful at avoiding the pressure from parasites in those years that precede the population explosions and defoliations. Could (summer) warmth play a role in this? Or is it the winter cold?

Investigations into these possibilities did not provide any clear correlation. The average summer temperatures at the start of the 1970s were below the average for the past 50 years. In the mass flight summer of 1974, it was almost 3°C lower than in 1983, the next year with an ermine moth outbreak. In 1991, as all the bird-cherries were once again defoliated and covered in webs, the temperature was more or less average, and in 1997 it fell to just one degree above the average for 1974. The temperatures for the winters preceding the mass flights were all within the average range, with only slight variations. In recent times, it has been no different. There is therefore no discernible relationship between the population explosions and the summer or winter temperatures. Rather, it appears that we are looking at longer-term frequency cycles with intervals from between 7–9 and 10 years. Another such cycle occurred in recent years – hence the media interest in the web-covered trees that did not die, contrary to expectations.

The new foliage that develops after defoliation is so good at its job that the years of defoliation can barely be identified in the annual growth rings of the tree trunks. On the contrary, one has the impression that the total leaf loss only corresponds to what the bird-cherries invest in the production of flowers and of fruit, those small black cherries, in years in which there is no ermine moth infestation. Naturally, these are destroyed in the event of defoliation. The second generation of leaves remains almost entirely undamaged and is thus highly efficient, thanks to the fact that the tree has no other insect species that can cause any noteworthy damage. A long-term effect is only imaginable, and even probable, if one follows the growth of the trees in the riparian woods under traditional coppicing.

Coppice management and its consequences

With coppice management, trees in riparian woods are cut back or 'coppiced' every 15–20 years: that is, their trunks are cut off quite close to the ground to leave 'stools'. The stools then put out new stems. When these are as thick as an arm or slightly thicker, they are felled, cut into lengths of approximately 30 centimetres, split and used for firewood. Coppicing is currently on the increase under the guise of renewable energy. Since 'renewable energy' is held to be fundamentally good, no consideration is given to the fact that burning such wood releases large amounts of tiny polluting particulates, significantly more than are released by diesel per unit of energy produced. The trees that have been coppiced – principally grey alder, *Alnus incana*, black poplar, *Populus nigra*, and, as it happens, bird-cherry – sprout again from the stools and over the years eventually produce compact groups of new trunks. Their inward bend just above the soil's surface will indicate this traditional style of management even decades later.

Grey alders should in reality have a major advantage when they grow because their roots have a symbiotic relationship with actinomycetes, thread-like bacteria. These bacteria can capture and bind atmospheric nitrogen directly and make this available to the alders. However, bird-cherries that have been coppiced re-sprout much more quickly. After a few years, they tower over the young alders and poplars. This height advantage will last until at least the end of the first decade of their growth. Only then will the alders catch up.

They will then overtake the bird-cherries, making them a secondary layer tree in the riparian woods, assuming that the woods continue to grow long enough for this to happen. This slowdown in growth could result from the hindering effect of an infestation of ermine moth caterpillars. However, this is barely more than an educated guess. It could also be the case that, after about a decade of growth, the root competition between the various tree species rises sharply. Yet there is a weak link in this argument that should actually be immediately apparent: why do the bird-cherry ermine moths not use the second generation of leaves as well, by developing another generation themselves? As can be seen, the new leaves develop in the best possible condition and are so abundant that it is clear that the defoliation that has taken place has not damaged the tree. A second generation of caterpillars in summer would fundamentally change everything that I have described and assumed. Why does it not occur?

Generations and multiyear cycles of ermine moths

If we try to imagine what takes place at the beginning of the first generation it will immediately become clear why a second is unlikely, if not quite impossible. As described, the tiny caterpillars that have been fully developed since the previous summer but have remained within the egg membrane hatch out at the same time that the buds of the bird cherry appear in spring. With their miniature jaws, they can only eat the very tender, very young small leaves that are also just developing. These are not available when the female ermine moth lays her eggs and places the clutches on the buds that are developing for the following year.

Egg-laying occurs from the end of June to the middle or end of July, depending on when the ermine moths emerged. It is too late to synchronize the hatching caterpillars with the newly sprouting buds in early summer. If defoliation has taken place, the trees affected develop new foliage much more quickly – indeed, as soon as the moth caterpillars pupate or just afterwards. Their development requires time, and at this point the tree is simply faster than the ermine moths.

The newly hatched caterpillars cannot feed on the new leaves since these have already become strong and tough, and maybe because they now contain too much of the toxins mentioned earlier. It is hard for the caterpillars to detoxify such toxins and the process would require energy that they do not have. Accordingly, the moth can only synchronize its lifecycle with the tree once, when the tree puts forth buds in spring. But the bird-cherry is not the only one to suffer from this kind of caterpillar infestation in its relationship with the ermine moth. Just how decisive this synchronization is, and what an important role it plays for numerous moth and butterfly species, above all for those known for their mass outbreaks and the damage that they cause, will concern us further when we consider the winter moth.

It is possible that there is a connection between the multiyear ermine moth cycles, which extend over 7–9 years, and the synchronization of the caterpillars with the tree. The fact that the cycles occur has been determined, but such determination offers no explanation for their occurrence. There must be a valid reason for them, but the compilation of records on an annual basis is far too coarse a tool to find out what it is. A schedule subdivided into weeks to document the flight times of the ermine moths already provides us with an instructive but quite unexpected finding. Mass occurrences, clearly indicated by measuring numbers of ermine moths flying into a light trap, only took place when the flight time in summer was quite normal, in the 28th and 29th week of the year – that is, early to mid-July. Mass flights are not generated by years characterized by above-average fine, warm weather in May and June, nor by years when there are cool and damp conditions at this time, which could affect the parasites, but rather by quite normal conditions. Weather as a direct cause of the multiyear cycles thus becomes even less likely. But then what can it be? The frequency patterns of some insects appear to correlate to sunspot cycles at various latitudes, not only in the polar regions, but these are significantly longer, with cycles of 9–11 years. Something else must have a decisive influence on the accumulation and subsequent deceleration of the population development of the ermine moths.

It would be appropriate to reconsider holding the parasitic wasps responsible, as the main predators of the caterpillars, since they cause the greatest losses when the caterpillar populations are of average to above average size. If their effect at this stage of the population growth is too small, then it is imaginable that their control will fail. The weather could therefore have a greater effect on the parasites than on the moths or their caterpillars. The fact that web-covered bird-cherry trees tend to become conspicuous in the whole of an area at the same time points in this direction: for example, in the whole of the northern pre-alpine region of Bavaria. Only weather could bring about such a widespread phenomenon.

Purely local effects, which do also occur, become meaningless on a large scale. If this plausible assumption is correct, then there must be connections between the population development of the ermine moths and special weather effects, which cannot be deduced from regional averages or even from annual ones. It could, for example, be short, sharp frosts in an otherwise average mild winter, or too much rain, or a period of dry, warm weather when the buds are emerging in spring. Weather patterns are complex. No year is quite like another, which is why the 'Hundred Year Calendar'* and similar long-term forecasts will always fail to measure up to reality. As necessary as they may be for meteorological statistics, averages are often meaningless when attempting to understand natural processes.

Parasitoids on other ermine moths

We can discern the strong influence of parasites by taking a comparative view of the webs created by the different types of ermine moth cited above. Without going into too much detail here, there are differences within a species, but there are even greater contrasts between the different species. Some, like the apple ermine, *Yponomeuta malinellus*, make only small, inconspicuous webs. If they have infested

---

* A book of weather forecasts compiled by Mauritius Knauer (1613–64), an abbot at a Cistercian monastery in Mannheim.

an apple tree and this has become noticeable, it is usually already too late for the blossoms. The caterpillars of this ermine moth behave in a very similar way to those of the bird-cherry. They hatch when the buds are put forth and consume the small leaves that are just unfolding, as well as the inflorescence.

In years when there is a heavy infestation, even if it does not lead to extensive leaf loss, the bird-cherry will therefore flower only weakly or not at all. The same is true of apple trees when they are infested by ermine moths, which have recently become quite rare. If there is a significant increase in the numbers of ermine moths, we will find the spindle-tree and the hawthorn bushes quite closely wrapped in webs too. This coincidence underlines the plausibility of special, weather-dependent facilitation. But when the caterpillars of the spindle ermine moths are ready to pupate, they do exactly the opposite of the bird-cherry ermine moths. They do not cluster together in thick masses that pupate collectively, but first weave a loose mesh with long, thin cavities, the size of thick dessert grapes. Then they stretch a thread through this, create their silver-white cocoons and attach them to the thread. The silk cocoons can hang in a horizontal, diagonal or almost vertical position. The individual pupae remain separate from each other so that the pupa web displays an interior structure that is fascinating to look through. The walls of the long cavities in which the pupae hang are thick enough to prevent the penetration of the parasitic wasps. They will try in vain to reach a pupa with their ovipositor, but the empty space between the web and the cocoon is too wide.

If a parasitic wasp should nevertheless manage to reach a pupa, then the web will give a little, just like a hammock suspended on an elastic rope. The special cradle-like structure presupposes a certain degree of parasitism. The fact that something so delicate came into being through the course of evolution indicates how significant the pressure of parasites on ermine moths was and still is. Nevertheless, and as so often, there is a qualification: this version of pupa protection has a major disadvantage. Such webs are far more susceptible to the effects of weather, such as heavy rain or gusts of wind, than the densely covered pupae masses of the bird-cherry ermine moth. As has already been emphasized, the weather is an abiotic factor that affects the ermine moths irrespective of the state of their population at the time; that is, it is density independent. It is probably because of this that the spindle ermine moths experience far smaller fluctuations in abundance than the bird-cherry ermine moths, and also do not normally undergo comparable outbreaks. *Yponomeuta cagnagella* thus behaves in a relatively normal manner in this respect, just like other moth, butterfly and insect species.

The lifecycles of butterflies and moths

Do we finally have the information we need to evaluate the massive changes in the occurrence and frequency of our Lepidoptera? A brief interim summary seems to be appropriate, so that the main features are not lost in all the details. Let us imagine, to this end, the lifecycle that all moths and butterflies go through. It begins with the newly laid eggs, in which the young caterpillars are developing. These hatch and feed on food that is suitable for them until they need to shed their skin in order to continue growing. After going through several such moults or instars, they are fully grown and have been prepared for pupation by means of hormones. For this purpose, the caterpillars seek out an appropriate place. Inside their bodies, the pupa then forms and frees itself, wriggling and pushing, from the final caterpillar skin. Some pupae hang freely, others lie in burrows, which were made earlier by the caterpillars; still others find themselves inside a web, a pupa cocoon, that protects them. The pupal period, a phase of complex internal transformation, can take more or less time depending on the surrounding temperature. Finally, the metamorphosis is complete, and the moth or butterfly can emerge. This is the reproductive stage (see Photos 23 and 24).

The nutritional intake of the moth or butterfly is limited to sucking nectar or taking in a few drops of water from dew or a puddle. Some species need mineral salts and certain other substances that they take from animal excrement or dead animals. But this is all extra: the actual feeding stage is the caterpillar. Its life therefore usually lasts significantly longer than that of the adult butterfly. The cycle is completed with the mating and egg-laying of the female. So far, so good, so simple: egg ⇒ caterpillar ⇒ pupa ⇒ butterfly ⇒ egg, etc.

Only nature is not simple, but highly complex.

The cycle must fit without interruption into the course of the year, even in the tropics where the seasons that are winter and summer for us are replaced with dry times and wet times that have a similarly strong influence on insect life cycles. Even in the permanently humid equatorial tropics, living conditions are never continuously the same. The cycles of the moths and butterflies (and all other living things) must adapt to cyclical or even irregular changes in nature. Where we live, in the temperate climatic latitudes, summer and winter form the two poles around which the lifecycles turn. Summer, more precisely, the growing season, promotes, while winter, especially frost, inhibits activity. Each new generation must pass through bottlenecks that the environment sets, and must overcome adverse periods, such as winter or great heat and drought. We are, in principle, familiar with this, even if as humans we have managed to significantly minimize our dependence on external conditions. Yet we are still affected by the two basic cycles of night and day and the course of the year, and accordingly heat our homes in winter and lengthen the days with artificial light. We attempt to impose our own notions and needs for sleep and wakefulness on the length of the day and the season of the year, with mixed success.

In nature, living things are far more subject to natural cycles than we are, but they do not need to surrender to them entirely. It is notable that, in the course of evolution, life has increasingly freed itself from the dictates of the inanimate environment and achieved extensive autonomy. We can refer to this process as the 'emancipation' of organisms from the dictates of the environment. It is wonderfully illustrated by the winter moth, which seems (to us) to have sidestepped the warm and comfortable summer world of the butterfly in favour of the transition periods into and out of winter, and a particularly delicate and fragile-seeming butterfly, which, thanks to its occurrence in woods and gardens, is still among the best-known of our butterflies, the brimstone *Gonepterix rhamni*.

Hardy Winter Moths

The vast majority of butterflies and moths fly in summer. The flight time of over 90 per cent of the Lepidoptera species is between April and September. They are most abundant from mid-June to the end of August. For the different species, the flight times of each generation only last a few weeks at the most. Many only have one such generation per year, some two or even three, but all together their occurrences overlap to produce the general pattern that we are familiar with in nature: a rapid increase in spring, with a peak around the summer solstice and a decrease towards late summer and autumn. The average figures for the flights of nocturnal Lepidoptera illustrate this sequence very clearly.

However, when we look at the whole year, we recognize some marked discrepancies. As early as the end of February, and in areas with mild winters even earlier, a first nocturnal moth will start its flight. Numbers will peak very prominently at the end of March and thereafter rapidly dwindle again. From the middle of April to the start of May, there can be several weeks when no moths are spotted at all. Only after that do the numbers start to increase rapidly and reach their climax, as I have just mentioned, in July. From mid-October, the numbers of moths counted increase again, producing a late autumn peak. The pre-spring and the late-autumn peaks do not generate nearly as large a moth count as that taken in mid-summer, but since they are quite pronounced and clearly set apart, there must be a special reason for these peaks. In fact, these numbers include species that are specially adapted to these unusual times of year: the spring and autumn noctuids and the spring and autumn geometrids. Both groups merit closer inspection; in particular the winter moths.

Winter moths belong to the geometer family (Geometridae)* and are linked to early and late frosts, as their German name, *Frostspanner*, expresses very appositely. They do not, however, resemble those Lepidoptera that are covered in thick hairs and are particularly protected against the cold. They are much more similar to the thin-bodied geometer moths that fly in summer. But it is this characteristic, that of flying after the first frosty nights of autumn when the temperature is just a few degrees over zero, that makes them so special. When they ghost through dusk as the season turns to winter and end up on the windshield of your car when you are driving at night, it is hard to believe that such a delicate type of moth can even move its wings under these external conditions. Hard to believe, too, that they actually require the first frosts as a signal that their time to fly has come. When they have become active, they do not just fly around alone, emerging instead in a regular swarm into the late twilight of cold, damp evenings in November or December. If your road leads through a beech wood and is not all that busy, then your headlights might well pick them up and betray their flight. They could easily be mistaken for small, dry leaves, swirled around by the wind, but they are alive and they are not aimlessly wandering around.

What they are trying to locate are the females of their species. Curiously enough, these are flightless. They sit on tree trunks like fat larvae, or crawl very slowly upwards until they have reached an appropriate height for the emission of their fragrance. This attracts the males. And since windless, cloudy and damp nights trigger the flight of the winter moth, the males must fly around in this erratic manner. This is the only way in which they can locate the scent-plume of the female and begin their more goal-oriented flight. If the females have

* From the Greek for 'ground' and 'measure', named for their looping style of locomotion.

already mated, they will cease to produce the attractant and merely crawl or drag their thick and heavy abdomens a bit further up the tree until they can deposit their eggs in cracks in the bark. From that point onwards, the eggs are left to themselves. After they have laid their eggs, the females visibly shrink like so many emptied sacks and soon die. The males also contract when they have used up the little energy reserves that they had for the flight through the cold night. There is no doubt that this is a peculiar way and time to mate. Triggered again by the mild nights following the winter frost as the winter turns to early spring, the same thing takes place with those of their relatives that are active early in the year, those winter moths that mate between the end of January and the beginning of March.

The spring species actually resemble the autumn species quite closely in their appearance, but the males can be differentiated by experts on the basis of the colouring, pattern and the shape of their wings. Since these are not present in the females, their differentiation is more difficult, but not problematic, since occurrences of the autumn and spring species are separated by winter. They are not easy to find during the day, even though the males sit on tree trunks with their wings spread flat but held at a slight angle, in a position that is characteristic of geometer moths. They are well camouflaged: their colouring is either a late autumnal brown or, for the earlier species that flies during the leaf fall, a dull rust-brown.

Life at the edge of winter

Why and how did these moths come to have such an unusual lifestyle? And why is this of interest to more than just lepidopterists?

The questions of why and how are very closely connected and, in order to answer them, we must look to the seasons. Late autumn, after the first frosts, and the time just before spring with its first mild nights are very distinct peripheral seasons in the lives of the moths and butterflies and a far cry from the optimal conditions that usually prevail in summer. One would think that a moth forced into the margins of the seasons in this manner ought to be rare. Such external conditions could hardly permit a high frequency of occurrence. Yet this assumption is clearly based too closely on our own feelings, since many of us are not fond of the dark transition from autumn to winter, that chilly and gloomy season that can make us feel slightly depressed. In colder climes, an early thaw in late January or early February can seem too early, since we know that winter weather is likely to return. But in fact, the frost-loving geometers belong to those sometimes very abundant species of moth that cause damage to orchards across regions. On the basis of their abundance, the living conditions for these moths must be classified as very favourable, and it will soon become apparent that they are. Yet they cannot be generally favourable, otherwise many more moths would fly at the same time as these geometers and not in the summer. The why and the how are therefore more complicated than originally thought – once again.

Let us attempt to understand and classify the life of a winter moth by starting with the answer to 'why?', a question that reaches further back into the past. Two reasons for its unusual lifecycle are apparent, and are similar to that of the brimstone butterfly, the butterfly known as the harbinger of spring due to its early flight. By the time of the first frosty nights, there are almost no insectivorous bird species with us in central Europe anymore. The tits that searched intently for caterpillars and butterflies of the right size earlier in the year have already switched over to a grain-based diet. Of wrens, which overwinter here and have not flown to the Mediterranean area, there is certainly no profusion, and, in any event, no birds will be out hunting insects in late twilight anymore. What about bats? By late autumn, they have already found their winter quarters. There they will hang, rigid, with body temperatures of just over 0°C. Even the large spiders in the woodlands and the gardens that could catch moths have stopped building webs. Early winter and very early spring nights are predator-free for the cold-loving geometers. They can whirl slowly through the air without worrying about any possible attacks. The fact that they would sometimes collide with cars was not foreseen during the long history of their development in the times before humans. But it also remains insignificant as far as their occurrence and abundance are concerned.

This enormous advantage does, however, have a hitch. No normal moth or butterfly can fly in temperatures of 2–4 degrees above freezing. The temperature is simply too low for this. The vast majority of species need 10 degrees more, at least. Favourable conditions for flight are external temperatures at which our naked human bodies are also in the so-called 'thermoneutral zone' – that is, approximately 27°C. In this zone we lose just as much body heat (without the cooling effect of sweat) as that which is produced by our metabolism when at rest. Our core temperature of 37°C will also remain stable with just the slightest energy input. If it becomes warmer outside, we need to sweat. If it gets colder, we need clothing or vigorous movement to provide additional warmth.

The same basic principle applies to the bodies of moths and butterflies, with the qualification that they are unable to maintain a stable, permanent body temperature. Like all insects and other animals, with the exception of mammals and birds, butterflies and moths are poikilotherms or 'cold-blooded'. At rest, or with only slight movement, their body temperature will correspond directly to that which prevails outside. No thick fur from scales or hairs will keep them warm. These can only slow down the flow of heat from the body if the heat is generated internally through the activity of the flight muscles. Only large moths and butterflies such as the hawk-moths can manage a kind of intermediate state with their efficient musculature and an intermittently high core body temperature. In this condition they resemble the hummingbird, not only in their hovering flight technique, but also in their extremely high internal body temperatures of over 40°C. This temporary production of warmth, approaching that of birds, is impossible for most moths and butterflies, because their wings are either too large or too small. The wings of hawk-moths are slender and 'streamlined' and adapted to their particular style of flight, whereas the wings of most butterflies and moths have a large surface area and are therefore excellent for soaring and gliding but not for flying at speed. The flight of many moths and butterflies therefore looks to us more like reeling and whirling.

The mastery of seasonal niches

All of this has much to do with the various species of winter moth. First, wings only develop in the males, as has already been noted. The females are wingless, except for the stumps, which are useless for flying. Second, they fly at temperatures at which their muscles ought not to be able to work at all. The flight muscles only operate properly from about 10°C, and only after some warming up, possibly supported by wing shivering. The fact that the males of the winter moth can do this at much lower temperatures is due to special enzymes, which effectively carry energy to the flight muscles from temperatures a few degrees over zero. Developing such enzymes was key in the evolution of the winter moths, and because these special substances are lacking in the majority of other moths, they cannot use the two seasonal 'windows' – the nights of the first frosts and the early mild nights in late winter – for their flight. At such temperatures, they remain as stiff as pokers.

The eerie flights of these moths on still nights at the edge of winter, which resound with the hoots of owls, therefore have nothing to do with magic. They are due to enzymes that function normally at such low temperatures. A description of the biochemical method by which this achievement is attained would be beyond the scope of this book. The third question, whether a consideration of the lifecycle of the winter moth is of more than merely lepidopteran interest, arises from the consequences of this special enzyme. The caterpillars can afflict fruit trees in spring to such an extent that they count as pests. In deciduous woods and parks, they cause such defoliation to trees and bushes in some years that they become unpleasantly disfigured. This visible success alone demonstrates that the winter moths' mastery or 'conquest' of these seasonal niches was worthwhile. It is likely that both groups, the autumn and the spring species, descend from ancestors that lived in regions that were largely frost-free in winter. With an increasingly cold (ice age) climate, snow and frost drove a metaphorical wedge between them, pushing the earlier and later forms further and further apart. I referred to caterpillars that hatched early and others that hatched late when we were considering the ermine moths on the bird-cherry.

As long as the spring weather fluctuates irregularly at the time when the bird-cherry trees come into bud, neither form of caterpillar will have an advantage over the long term; nor can they be differentiated from one another as either early or late forms. However, if the separation occurs at the time of reproduction, then, over the course of time, separate early and late forms will develop. Our winters have not always been the way that they are now. This is not the first time that the climate has changed, and it has not changed only because of human activity. We use the term 'climate change' to refer to long-term, lasting changes in the weather. Since nature has never been 'stable', these weather changes form part of the normal events that lead to the development of lifestyles. This is exactly why nature is so species rich.

Why female winter moths do not need wings

But to return to the repercussions and the phenomenon of the wingless or stump-winged females. As already described, they are little more than egg-sacs with limited mobility. Following successful mating and fertilization of the eggs, they lay several clusters, although these do not have to be located in specific places like those of the ermine moths. Cracks in the tree bark are quite sufficient. The eggs are frost-hardy, like many moth and butterfly eggs. If the increasing temperatures in spring add up to a specific heat accumulation, measured in 'degree days' by forest entomologists, then the tiny caterpillars will hatch and set out, equipped with a hair-thin thread, in search of a bud that is also just emerging. The buds react to a similar sum of 'degree days' and to the length of the day. What happens next corresponds to the emerging caterpillars of the bird-cherry ermine moth, with the significant difference that the latter are tied to the bird-cherry, while the winter moth caterpillars are not at all particular, instead feeding on numerous different trees. Manfred Koch's book *Wir bestimmen Schmetterlinge* [*We Identify Butterflies and Moths*], which for me has been the most important field guide for native moth and butterfly species and the one I use most, remarks of the caterpillars of the common winter moth, *Operophtera brumata*, that they feed on fruit trees (apple, pear and plum) in preference to nearly all broad-leaved trees and that they are an 'almost annually occurring, serious fruit tree pest'. The mottled umber, *Erannis defoliaria*, also infests a wide spectrum of broad-leaved trees and is only less feared because the affected trees are supposed to deliver wood in the long term, rather than producing fruit every year. The same is true of the northern winter moth, *Operophtera fagata*, which appears in masses in some years. Its caterpillars prefer to feed on beeches, but also feed on birch. The dotted border, *Erannis marginaria*, is becoming less common. Evidently, late flight and reproduction after the first frosty nights is a more successful form of adaptation than the very early flights in early spring. The numbers of geometer moths that were recorded using UV lights in autumn was about four times as high as the numbers of geometer moths in spring. Overwintering as eggs probably incurs lower rates of loss than hibernating as pupae that must emerge in the first mild nights of early spring. The large number of eggs and the survival of the clusters are therefore key factors in the life of the geometer moth. If the caterpillars are not specialized on a particular food plant species, the clusters of eggs can be hidden in the cracks of the bark. Pupae are more vulnerable than eggs, particularly to damp and fungal infection in especially mild winters. The special significance of the large egg count underlines the phenomenon of the females not developing wings. All the material that would be necessary for this and the flight muscles is instead used for the development of eggs. Wings are not needed; it is enough if the males fly to the females. The quantity of sperm cells produced by the males is insignificant compared to the masses of eggs and requires no additional material that could be saved.

Deforestation, poison and the decline of the codling moth and the winter moth

Bird conservationists might say that it is good that there are winter moths because the caterpillars of these moths tend to make up by far the most important food for the nestlings of numerous bird species, as well as their parents. The caterpillars contain no toxins that would be dangerous for the birds and are available almost every year at the right time. Only the small green caterpillars of the green oak tortrix, *Tortrix viridana*, which sometimes cause defoliation of oaks, are as important. The fact that in some recent years they have eaten the oak leaves earlier than usual, at a time when the pied flycatcher and some other bird species are not yet ready with their broods, is considered a dangerous development because it alters a closely linked network of relationships in nature. Quite how far the effects of airborne fertilizers, which all our forests have been exposed to for decades, are involved has not been considered. However, overfertilization is quite a different sort of factor from that of the climate and is conceivably more likely to affect growth than either an increase or a decrease in air temperature by a single degree. On the other hand, winter weather should have a much stronger influence on the winter moths, since they fly, as described, at very specific temperatures and at high levels of air humidity, during much narrower time windows at the beginning and end of winter. Since the 1960s, their numbers have drastically declined. However, not all species have been affected equally: the impact has primarily been on the common winter moth, *Operophtera brumata*. In Germany, the numbers counted in the years 2013 to 2017 were about 80 per cent lower than those for the 1970s. Is this because of increasingly warm winters? Since the northern winter moth, a typical woodland and woodland-edge species, and also the mottled umber, *Erannis defoliaria*, have not only not declined but, instead, significantly increased in number, the effect of climate change on the winter moth is doubtful. After all, changed winter temperatures affect all species of geometer moth.

A similarly striking decline in a quite different moth species from the leaf roller family (Tortricidae) points to the real reason. I am referring to the codling moth, *Cydia pomonella*. Its caterpillars were generally known as the 'worm in the apple' when apples still came direct from trees and not via supermarkets (pre-sorted and selected meticulously and precisely in accordance with EU standards). The current population in Germany amounts to only about 12 per cent of that recorded in the 1970s, even though the codling moth can also live on other species of fruit and nut tree and not just apples. The simple reason for the sharp decline in the numbers of both is the drastic reduction in the quantity of apple trees.

Grubbing-up premiums, offered by the state, led to the destruction of many orchards and meadow orchards 50 years ago in Germany and in large areas of the EU. Fresh fruit is now produced as far as possible in uniform, standard sizes in fruit tree plantations. Codling moths and winter moths are controlled using insecticides. The mottled umber, which lives on hornbeams, etc., in hedges and at the edges of woodlands, is not affected by the treatment, even if mechanized hedgerow care frequently seems to verge on hedge destruction.

The common quaker moth in early spring

Finally, I would like to have a brief look at the other moths that contribute to the peaks in numbers that occur in (late) autumn and in spring. As I emphasized at the start of this chapter, the noctuid moths belong to these. In spring it is the so-called common Quaker, *Orthosia cerasi*, which flies in great numbers shortly after the equinox and is also to be seen during the day on flowering willow catkins. More than 100 such common quakers can fly into a light trap in a single night. It belongs to a group made up of a handful of species that also includes the powdered quaker, *Orthosia gracilis*. Their bodies are covered with quite dense, furry scales, which clearly protect them on spring nights when the weather suddenly becomes cold and frosty. The yellow-line quaker, *Agrochola macilenta*, is similar, with its colouring and the patterns on the upper sides of the wings resembling the yellow and rust-brown of autumn, camouflaging it in fallen leaves during the daytime.

A few species of noctuid moth even overwinter as moths. Occasionally they will fly in the middle of winter, on mild nights, and we will see them sitting on shop windows where they have been drawn to the light. As a group, they are less uniform than the frost-loving geometer moths, but with the extension of their flights into autumn and the start of winter and with the early flight of the common quaker, they show the same tendencies. They both use the seasonal peripheries when there is little predation pressure, and when the most significant predators of the larger moths, that is, the bats, are already or are still in hibernation in temperate continental areas. The reaction of bats to outside temperatures is similar to that of most moths: an early cold snap in autumn drives them to their winter quarters, while early warmth in spring will wake them up. But everything remains within the relatively flexible limits of the seasons. The changes to the length of day are better triggers than the changing seasons. They guarantee that the variations in the weather are not too misleading.

Too precise a synchronization between species and seasons that produced too rigid a dependence would be overly risky. It is questionable whether any species of the middle geographical latitudes, in the large transition zone between variable Atlantic weather in the west and more stable continental weather in the east, has developed such a tight time constraint that it cannot adapt to the fluctuations of the weather. Put more simply: an early spring is early for the whole of nature, just like a longer autumn is long or an early winter cold snap is too early. The assumption that only certain species react to shifts in weather patterns cannot be fundamentally excluded, but is anything but likely.

In magnitude, the temperature ranges greatly exceed the fluctuations between years and the average temperature trends (if there are any) and temperatures in a single day can differ by 20°C or more.

Brimstones:
The First Spring Butterflies

Brimstone, small tortoiseshell, peacock or red admiral: which is the first to be seen in spring? In Bavaria, we usually see a brimstone first of all: as a rule, a male. A warm day with a *föhn* wind at the end of February is all it takes, and already they are in flight. In the woods along and near the River Inn, close to where the Salzach joins it, I spotted brimstones first five times between 2011 and 2016. Only in 2013 did I see a small tortoiseshell first, a couple of days earlier. But perhaps it is the peacock that we think of as being the first true herald of spring – as it flutters and tries in vain to get out through the window of a garden shed, we feel the need to rescue it. Which butterfly we see first actually depends on the type of weather experienced in the run-up to spring. If you want to delight in their delicate beauty, look out for spring butterflies as soon as it is sunny and warmer than 10°C.

It is the warmth that wakes them up, since all these early butterflies spend the winter in their butterfly form. In this sense, the brimstones are particularly tough. They find a place to suspend themselves in autumn, somewhere in the undergrowth in the wood or in a dense bush in the garden, and fall into a winter 'torpor'. A substance like antifreeze in their bodies prevents them from freezing (see Photos 25 and 26). Brimstones can survive at minus 20°C, although for peacocks and small tortoiseshells this temperature would be fatal. The two latter species must find protected areas such as garden sheds, old cellars or other nooks and crannies, and such places are rare. For this reason, far fewer peacocks and small tortoiseshells spend the winter in central Europe than brimstones. The lovely red admiral manages it only very occasionally in Germany, when the winter turns out to be very mild. The comma, *Polygonia c-album*, survives the winter far better. This butterfly, recognized by the small white 'C' on the underside of its hindwing, spends the winter as a butterfly and is similar to the small tortoiseshell, but can be differentiated from it by its very jagged wing edges. I dealt with both of them as members of the group of nettle-feeding butterflies in an earlier chapter when I discussed their caterpillars and their flight to central Europe from the south in spring (pp. 47 et seq.). They fly there very purposefully and warm themselves up every now and then with short rests on the ground if the air is still too cool. It appears that the wings work as a kind of solar panel for them. The brimstones do not fly like this, sunning themselves rarely if at all.

The bright lemon-yellow brimstone males dance around at the edges of woods and along forestry paths or garden fences as if on established routes. First they fly in one direction, and then they turn around and flutter back in the opposite direction, sometimes along the very same trajectory. Evidently this is how they monitor a territory. They cover the same section regularly, as soon as the day is warm and sunny enough. Males that do not yet have a territory search for one that is not already occupied by another lemon-yellow butterfly. Now and then there are confrontations in which the males whirl around each other until one of them suddenly flies away. Whether this was the original occupant of the territory or the intruder will be unclear without close inspection of their markings.

It is quite unusual for a butterfly to defend a 'territory'. Given that the lives of butterflies are so short, one would have thought this type of behaviour was not profitable, since it requires energy. Indeed, there are only very few butterflies outside the tropics and subtropics that display territorial behaviour. It is only worthwhile if they live long enough and if the females of their species do not all emerge at once but over a longer period of time, or indeed, require time until they are ready for mating, as is the case with the female brimstones. Even for temperate zone butterflies, brimstones are extraordinarily long-lived, with lifespans of up to 10 months. Tropical butterflies mostly do not achieve such longevity. Admittedly, brimstones rest for two-thirds of their lives, but the time they spend resting is still several times longer than the average active life of other butterflies.

Butterfly lives are simply too short. One of the few other exceptions that we find in Europe is a medium-sized brown butterfly with white spots, the speckled wood *Pararge aegeria*. Rather like the brimstones, but later in the year and across two generations, the males of this widespread, but not particularly conspicuous, butterfly patrol woodland trails, lanes and the like in broad-leaved woodlands and forests and attempt to keep intruders of their own species out. Neighbouring males frequently indulge in territorial encounters, spiralling around each other until one of them retreats. However, their ability to recognize their own species is often not very accurate, and they sometimes fly at other butterflies that only superficially resemble them. They mostly do this from a vantage point that offers them a good perspective. They even manage to displace the much larger peacock successfully. I once had the impression that a speckled wood wanted to banish me too: when I got too close, it flew right in front of my face! When I retreated, it turned around and landed on its seat again.

Butterfly attacks

Even more impressive was a 'butterfly attack' that I experienced once in early summer on Istria. I was walking along a narrow path that led through the dense *macchie* shrubland to the beach, when all of a sudden, a sinister fist-sized 'eye' flew directly into my face, causing me instinctively to draw back. At first, I did not understand what had happened and looked around in consternation as if I had had a hallucination. I saw nothing but a large butterfly fluttering about. It was late morning. The sun had not yet reached its full strength, but it was already so high that all the shadows were shortened. Assuming that I had just imagined it, which would have been peculiar enough, I went a couple of steps further. Then it came again, the 'fist-sized eye'. Now I realized what was going on. It was a two-tailed pasha, *Charaxes jasius*, an almost tropical beauty, whose relatives really do live in the tropics. It flew at me, and in such a targeted manner that the ochre borders of its wings resembled the edges of a giant eye. Fist-sized is barely an exaggeration. It stopped less than a handspan from my face and turned sideways, just to fly at me again a few seconds later. No doubt about it, the huge butterfly was trying to drive me away. Naturally, I let it win and drew far enough away in order to observe it more closely – through my binoculars. Just like the familiar little speckled woods in the riverside forests of south-east Bavaria, it took up a post that enabled it to survey the path over a total distance of more than 10 metres. When a red admiral flew along the path shortly afterwards, the pasha drove it off in seconds.

It goes without saying that this behaviour is all about females. The brimstone males patrol their stretches for two to three weeks before the first females can be seen. These are very distinct from the males, with their paler yellow, which is quite similar to the colour of the large white. This is probably why they are often not even recognized as brimstones. In any event, the males patrol for a surprising length of time before the females even fly. In the warm spring of 2014, in which there was a short cold period at the start of April, they were active for almost a whole month in advance. In 2013, there was heavy wintry weather in March that lasted until April. In that year, the flight times of the males and females were crammed together into April and the first half of May.

The problem with early flight

Why do the brimstones fly so early, given that they could easily wait until May when conditions would be more favourable? They could save themselves the long period from the end of February to the middle of April, when the weather is unstable and fluctuates between premature warmth and delayed winter cold. April frosts are not rare; sometimes there are frosts until the middle of May in Bavaria. And yet, like the spring and autumn geometer moths, the early brimstones have the advantage that most of the insectivorous songbirds only return from their winter homes between mid-April and early May. Even if the songbirds saw them, these early butterflies would not be worth chasing. To be worth pursuing, the quantity of insects must be sufficiently large both for the parent birds' own nutrition and for their young, especially when their first brood has hatched. Between the end of February and the middle of March, the useable quantity of insects is nowhere near sufficient for this. Brimstones are highly visible though. For skilful hunting birds, they should be easy prey. Even starlings sometimes manage to snatch a brimstone, usually a male. The females are more usually overlooked, since, as observed above, they strongly resemble the cabbage whites, especially the female large whites. These are also protected by toxic glucosinolates, as explained above (see p. 76).

Brimstones could therefore be mimics of the poisonous cabbage whites. Or, if they contain distasteful substances from the principal food plant of their caterpillars, the alder buckthorn, *Frangula alnus*, which they might, perhaps they are part of a mimicry-ring of butterflies that signal that they are inedible by means of their conspicuousness. This is especially important for the brimstones in summer, when they have newly emerged and must survive the weeks and months before their hibernation begins. Let us therefore have a closer look at the males (lemon-yellow and very conspicuous) and the females (pale yellow, almost white). Should they not both need the same level of protection?

It is possible that they receive it. After all, what seems white to our visual faculty might be closer to yellow for birds, since their colour sensitivity is so differently constructed from ours and extends into the ultraviolet range. Which means that the difference between males and females is perhaps greater to our eyes than it is to the visual perception of birds. Whatever it may be, research will be able to clarify it with the possibilities open to us today. The difference has, at any rate, a seasonal dimension. The males fly markedly earlier than the females in spring, but at the same time as them in the 'summer generation'. They only appear to produce a new generation, however, since the 'new' summer brimstones are in fact the same ones that will fly in the following spring. They will have a short summer rest and, following a short flight time in autumn, will seek out a place where they can spend the winter. In the summer, one is hardly aware of how many brimstones there are, since they fly in such small numbers and so inconspicuously. This makes them seem fewer than they really are. They should be rare, or at least inconspicuous, since the rule for imitators that engage in mimicry is that their model must be significantly more numerous than they are. The reason for this is obvious: the birds, which are supposed to be deterred from catching toxic (or bad-tasting) butterflies, cannot be allowed to have too many positive experiences in case they build up a 'search image' for those butterflies that taste good after all. That would mean that their protection had disappeared, that their distinctive features were no longer a saving grace but, instead, a deadly disadvantage.

Müllerian mimicry

Obviously, the activity levels of insectivorous birds are incomparably higher in early summer than they are in spring. Brimstones do not mate immediately after emerging, because their reproductive glands only start to mature as the days lengthen in the spring. There must therefore be an advantage to early flight. The brimstones do then mature, with the help of the sun. It is safe for them to expose themselves to it since there are no songbirds about that would catch them. Once the females are ready and flying in search of suitable places to lay their eggs, the cabbage whites are already in flight, and indeed in much greater numbers than the brimstones. The females can then systematically lay individual eggs one at a time on tiny alder buckthorn leaves that are just budding: usually just one or a few per shoot. That is where the caterpillars of the brimstone butterfly will feed. Since the alder buckthorn also contains substances that have a toxic effect, a certain level of toxicity ought also to occur in the brimstones. However, the anthraquinone glycosides from the alder buckthorn leaves are not nearly as toxic as the glucosinolates that the cabbage whites ingest. At least, evidently not for the birds, since the stone fruits of the alder buckthorn, which are poisonous to us, are very readily eaten by birds, which then spread the seeds that they contain. This means that the relationship between the brimstones and the (large) cabbage white is actually based on a kind of intermediate form of the two different types of mimicry. The more subtle type, known as Müllerian mimicry, is named after the person who discovered it, the German Fritz Müller, who was living in southern Brazil at the time, and is distinct from the type that was discovered earlier, Batesian mimicry, named after the American Amazon researcher Henry Bates. Batesian mimicry refers to the simple imitation of a poisonous or inedible species by one that is nontoxic and edible. In contrast, Müllerian mimicry includes groups of species that are similar to one another and that are all toxic or inedible to varying degrees. The advantage of this type of mimicry is that it does not matter how numerous the participants are because their abundance does not diminish the effect of the mimicry. This, for example, describes the relationship between our various species of wasps. They are all characterized and protected by the yellow and black 'wasp livery', and all of them have a poison sting with which they defend themselves against attacks by their enemies. The fact that specialists such as the honey buzzard, *Pernis apivorus*, are not deterred by this ultimately confirms that nothing in nature is absolutely certain. Our own spontaneous reaction to wasps, on the other hand, demonstrates clearly enough that Müllerian mimicry works very well. We do not first look to see whether it is a common wasp, a German wasp or a harmless hoverfly flying towards us before we get out of its way.

The brimstones merit this excursion into mimicry relationships since their example once more underlines how difficult, if not impossible, it is to try to force nature into strict rules and categories. Anyone who insists on 'definitions' and attempts to lay these down like laws has not understood nature. Brimstones are true imitators of poisonous models, since the great similarity of their females to the large whites works in this manner. But taken together – males, females and large whites – they also make up a Müllerian mimicry ring. This is composed of various different poisonous and bitter-tasting butterflies. As mentioned above, the anthraquinone glycosides do not offer as secure a protection as the glucosinolates of the cabbage whites. And yet, the brimstones can become quite numerous compared to the cabbage whites, particularly in spring. If they were pure Batesian mimics, then they would be hunted in a targeted fashion just at the most important time, when the females lay their eggs. Although this alone already makes them remarkable butterflies, they also need to be in a position in the spring to react flexibly to the cold snaps that will most likely come in March, April or even May. Being flexible here means being able to postpone their activity at short notice by several weeks. The spring seasons of 2013 and 2018 necessitated just such marked displacements, with May 2013 not being particularly favourable and May 2018 being extremely favourable to the brimstones. But what were the consequences? How was it possible for particularly good numbers to be recorded in spring 2014 despite the extremely unfavourable weather conditions experienced in spring 2013?

Once again, we are presented with the same questions. How does weather affect our butterflies? Does the winter weather influence the survival of the brimstones? The spring numbers alone would not provide sufficient information, since after all it is not possible for more butterflies than were alive in autumn to survive the winter. The population in the preceding summer must therefore be taken into account in an evaluation of the effect of winter. The spring records that differ from year to year could depend either on the population size of the preceding year or on the survival through winter – or both.

The critical factor of spring weather

In 2014, more than four times as many brimstones were recorded as in 2013, a fact that could be due to heavy losses resulting from the poor spring weather of 2013. However, when I consider the abundance of brimstones over the years, I do not think this can be the case. The extremely cold February of 2012, with weeks of night frost and temperatures as low as minus 20°C, did not harm the brimstones all that much. The halving of numbers in spring 2012 as compared to spring 2011 was (also) connected to the fact that the summer numbers in 2011 were already rather low. There were thus far fewer brimstones going into hibernation in the autumn than in the preceding year or the year after, summer 2013, which was followed by 'record numbers' in spring 2014.

Contrary to what one might expect, the conspicuously high spring population of 2014 did not lead to a higher summer population, and the spring butterfly numbers for 2015 fell to less than a third of those for 2014. Nevertheless, the summer population of 2015 then grew, and we eventually had another relatively good early spring flight in 2016. Confusing enough? That is just how nature's cycles are! All the weather phases of the year affect each generation of butterflies; a dry and hot or rainy summer, a changeable autumn, a cold or mild winter, spring weather for the first flight and finally early summer, when the caterpillars grow, pupate and emerge as butterflies. A desire to focus on 'winter' in the lifecycle of the brimstone butterfly or even to hold any changes to the winter weather (key term: global warming) responsible for the population growth and future of these butterflies would only be a sign of ignorance. Long-term research projects demonstrate this. They also remind us that counts can be unreliable, since one can never record moths and butterflies in exactly the same manner over several consecutive years. Many samples are required for a successful count, in order to recognize a trend spanning several years, if one exists at all. Fluctuations from year to year are misleading. They often happen to create increasing or decreasing data series that become insignificant when we consider a longer time-span.

The only thing that is certain is that the key factor for the mid- and long-term population growth of the brimstone butterfly is reproductive success in spring and early summer. The development of their populations is far less dependent on how they survive the winter. This conclusion sits most comfortably with the wide geographical distribution of these 'white' or Pieridae butterflies, which includes every kind of winter climate from the mild Atlantic west to the continental east in Europe. A fine spring therefore provides the best conditions for the brimstone butterfly, and we can verify this with our own eyes. This categorization also fits in well with the occurrence of its close relative, the Cleopatra butterfly, *Gonepteryx cleopatra*, to the south of the Alps. This brimstone butterfly is adorned with an orange-red area on each forewing and is slightly larger than our kind. In the Mediterranean summer, the Cleopatra experiences high temperatures that would be difficult for our brimstone butterfly to survive. The summer period of rest or 'aestivation', which has already been discussed, taxes the bodies of these delicately constructed butterflies far more than having to endure frost in winter. This is because heat draws moisture out of the body far faster than a dehydrating frost. The storage of glycol, the anti-freeze agent, can help prevent dehydration due to frost because it can prevent the formation of ice crystals within the body of the butterfly down to temperatures as low as minus 20°C, and can hinder excessive water loss that would cause the body to become dried out.

We can therefore well imagine that the ability to survive the summer heat and the dehydration that it caused was the prerequisite for the brimstone butterfly surviving the winter in the open. Only very few butterflies manage this. Butterfly specialist Eberhard Pfeuffer, who has focused on the brimstone butterfly in particular detail, even observed that, in winter, when the snow begins to melt, they creep a little way towards the surface, out of the ground cover into which they have retreated to overwinter. Damp, cold winter weather is less favourable for many overwintering insects than dry cold, partly because the damp also encourages the growth of fungal hyphae that can attack and kill pupae and butterflies. The parallels to human beings are once again clear: dry and cold conditions cause fewer difficulties than wet and cold weather. Frost is also more favourable in light of the typical communicable diseases that appear in winter, such as the various forms of flu or colds. To generalize, mild winters are not better 'for nature' than cold ones. Therefore, it may not simply be assumed that they will favour those species that require warmth. Nor may it be assumed that many species would profit from global warming, which for us particularly affects the winter, to such an extent that they would 'get out of hand' or because nature will somehow 'lose its balance'.

'Balance' in nature

How do these examples and digressions assist us in understanding which factors regulate nature? What provides that mysterious 'control' that, according to conventional wisdom, ensures that insects do not inundate and ruin our world with their powerful reproductive potential? The mere fact that this does not happen gives rise to another fact of nature: that favourable conditions are never permanent anywhere. This fundamentally excludes the excessive growth of any single species. Over the short or long term, and usually even after a very short period, conditions deteriorate and the growth of a population ceases. This is the effect of density-independent regulation, which has already been mentioned. The sequence of favourable and unfavourable external conditions fluctuates within and between years, and such fluctuation is more or less (statistically) random. This results in similarly more or less random fluctuations in populations. Population stability, in the sense of the same number of butterflies of a species or even every species existing in a given area, year on year, is nothing more than wishful thinking. We can produce such apparent lack of change if we convert the numerical changes to logarithms. This mathematical method reduces high or low swings to such an extent, in particular if the so-called 'common logarithms' are used, that the now only slightly fluctuating figures average out to appear as a straight line. Whoever wishes to, may see in this the 'balance of nature'.

However, in real-life situations it means little, if anything, if the purpose is, for example, to avoid economic losses. This can be illustrated with a numerical example. If the population of a species of insect that can harm agricultural crops grows, in the fields or in the woods, by a factor of 10, 100 or 1,000, this is represented on the logarithmic scale by the factors 2, 3 and 4 (the powers to which 10 is raised, to represent increases of those amounts). This compression of large figures to a scale that is easy to grasp is helpful for graphic representation and mathematical analyses, but seldom outside in the fields or in the forest. Where abundance is decreasing, and the purpose is to protect a local or regional species from disappearance, such a logarithmic conversion is even less useful. If the 'balance of nature' is justified in this manner, then the actual imbalances that sustain life and its momentum disappear. If everything were to stay the same, then the same conditions would persist, and evolution could not take place. We should liberate ourselves from the notion that everything in nature would follow its true course and all would be 'fine' if only humans, their actions and works did not exist.

Let us consider density-dependent effects once again, bearing this warning in mind. They come about not only because of the influence of natural enemies, parasites or pathogens, which together act more powerfully to curb population growth the larger the population has become; this is exactly what the description 'density-dependent' means. The availability of food also provides regulation in a density-dependent manner. If food becomes scarce, then the organisms will not survive or will suffer a greatly reduced fertility, as was observed with the mass reproduction of the bird-cherry ermine moths. Simplifying the situation, we could imagine that the species and its population – for example the caterpillars of a species of butterfly – exert 'pressure' on the availability of its food plant, while it is itself exposed to the pressure exerted by its predators and competitors. Clearly, the usage rate of the food plant and the rate of loss to predators must offset each other at least in the medium term, so that the overall balance does not swing too far either one way or the other. If it did, this would mean that either the population would decline and disappear over the short or long term, or that it would show sustained growth and thus become unstable. The model is once again simple and clear: reproduction rate (b) minus loss rate (m) ≈ 0.

An equilibrium of this kind would signify (biological) stability and constitute the zero-growth rate that is so often heralded as the prerequisite for humanity's survival. Yet once again, nature is more complex than the model. It is not a closed room with uniform conditions, in which the relationship between one species and another or between a small number of species is played out. Instead, there are hundreds, thousands, indeed thousands upon thousands, of species involved, depending on how narrowly or broadly we define the scope of the area to be measured or the spectrum of species to be considered. As soon as we transfer our model representations into the natural world, a multitude of diverse local conditions called the 'biotope' forms the basis for study. To this we must add two further extremely important processes, which were not taken into account in the explanations set out above, namely, the migration of regional excess numbers ('overproduction') to other places and areas, and its opposite, the immigration from outside into the places under consideration.

This characterization is closer to reality. The species are not all distributed equally: on the contrary, their distribution is more like a mosaic. In every place where they occur, they live in populations that fluctuate at varying rates over the years. In short, there is a constant state of flux in nature, just like the rising bubbles in a pot of boiling water. Increases and resettlements in one place can signify continuous general developments but may equally compensate for losses and extinctions in another. Although a more detailed consideration would lead to the central concepts and models of population ecology, which are outside the scope of this book, certain key conclusions can be drawn:

1   In order to be able to differentiate continuing trends from natural fluctuations, the time periods used for the investigations must be long enough. The question of how long is 'long enough' will be evident from the length of the fluctuations. There must be a sufficient number of years or generations available in the recording period to enable fluctuations to be distinguished from the main trend, if there is one.

2 Local investigations must be representative of larger areas in order to acquire significance. In a newly developed, convenient 'biotope', such as those that are now regularly established as compensation for encroachments into nature and landscape that have already taken place, developments that seem markedly positive may take place over many years (from the perspective of species protection), while on a larger scale (outside the biotope), the trend is for a sharp decline and is thus negative.

3 Comparative trends must be independent of one another, in the same way that weather and climate trends are independent from butterfly populations, if they are ultimately to be taken into consideration.

4 Recording methods should (or rather must) not be influenced by the personal abilities of the people that undertake the investigations. Physical and automatic measurements are clearly preferable to 'mere observations'.

5 Any comparisons must be reasonably justified when interpreting findings and drawing conclusions.

These five central criteria are usually based on the assumption that suitably detailed species knowledge is available. Many moth and butterfly species are not easy to identify; some cannot be determined, or can only be determined insufficiently, without special laboratory investigations. Other groups of insects demand even better knowledge of the species. However, specialists are rare, and their time is limited; their readiness to spend a lot of time identifying difficult species is limited too. But this should not discourage the willing from 'counting butterflies' and contributing to the knowledge surrounding the occurrence, abundance and population trends of butterflies and moths. The methods dealt with here are tested and practicable. Butterfly and moth counts are needed at home and abroad, since many species are under threat and in the process of disappearing. Those that are still numerous may soon suffer the fate of those whose numbers have already been decimated.

The examples that have been discussed so far may not have given the impression that moths and butterflies fare especially badly in the modern world. Yet it was important to gain an understanding of the processes that take place in nature with reference to the species that are well known and still numerous, and to demonstrate how diverse the mechanisms of cause and effect are. Even if in many cases it is an irrefutable fact, placing their extinction at the beginning would probably not have been the best way to inspire an interest in and an enthusiasm for moths and butterflies. That, however, was my intent. In the second part of this book we will consider the disappearance of the moths and butterflies and what can (still) be done to prevent it.

Part II

The Disappearance of Lepidoptera

Assessing the Abundance and Occurrence of Butterflies: A Major Challenge

Butterflies are here one moment and gone the next. They flit in out of nowhere, settle briefly on a flower, evidently drinking nectar, then try the next flower or fly on. The flight of some butterflies could give the impression that they are slightly tipsy. Nocturnal Lepidoptera, on the other hand, such as the hummingbird hawk-moth, which often also flies during the day, appear as swiftly as an arrow, hover in the air before their chosen flower, and then disappear again. This small hawk-moth beats its wings at such a frequency that they appear as a shadow to us.

As we amble through the grass in the garden, provided it is not as short as an artificial lawn, we sometimes disturb small, long-bodied moths that move out of our way, half hopping, half flying. This group, Crambinae, is known as 'grass-moths'. They belong to the exceptionally diverse grouping of 'small moths', the Microlepidoptera, which includes numerous different families. In fact, this label is generally used to indicate all those that do not belong to the families of butterflies, hawk-moths, tiger moths, eggar moths (Lasiocampidae), noctuid moths, geometer moths or a few other small but well-known moth groups. The naked eye can barely recognize the tiniest of these smaller families as being moths. In contrast, the largest hawk-moths are more like small birds, such as the hummingbird, with respect to their body mass and flight patterns. In fact, a similar level of energy conversion takes place in their bodies (see Photo 27).

How is it possible to establish the 'abundance' of moths and butterflies under these circumstances? Is it only possible if recording is limited to a single species or certain well-defined groups of species, for example 'butterflies' or, within these, the 'cabbage whites'? The latter are at least easy to spot, and usually they are not in much of a hurry. For peacocks or swallowtails, too, one can imagine being able to count them in specific areas or along (meaningfully) defined transects. Some examples will be discussed below. They show that something truly important can be drawn from transect counts. But moths and butterflies as a whole somehow elude such direct methods of counting based on sight and over defined transects.

Those responsible for the decline of moths and butterflies often dismiss the claims made by older experts and nature lovers that there used to be many more than there are today, saying that these are nothing but fanciful imaginings. Do they have a point? We know that our ideas about the 'good old days' are usually just nostalgic glorification. Moreover, our landscapes have changed and will continue to change. This is the passage of time that develops into what we call history. Nevertheless, seasoned experts and nature lovers are right. There used to be many more butterflies, moths and other insects and we can be relatively certain when things started to change. The major change began with the massive (over)fertilization of, and the introduction of pesticides to, agricultural land, and with the land consolidation that followed the Second World War. The second half of the twentieth century saw a great transformation in the way land was managed. Present conditions are the extrapolation of these events, significantly exacerbated by the repurposing of massive agricultural areas to produce biodiesel and biogas. The revolution in renewable energy represents the most recent stage in the industrialization of agriculture. I intend to prove this claim.

At the beginning of Part I, I described the method by which results were obtained that evidenced the stark decline in moths and butterflies. This procedure will now be explained in more detail. Since the investigations were carried out over a long period of time, these research activities led to findings that would not have been remotely foreseeable at their commencement. At that time, in the late 1960s, counting moths and butterflies hardly seemed important enough to qualify as university research. Even after 10 years, what would become manifest after 50 years was barely starting to emerge. In this respect, the research undertaken over the first two decades was not concerned with providing information about the changes to the occurrence of nocturnal moths and other insects. At the beginning, it was just curiosity: I wanted to see what was flying around on the outskirts of the village, in the riparian woods and later, much later, in the city. I was taken with the abundance of forms and had succumbed to a fascination with the new, like so many collectors: hawk-moths and noctuid and geometer moths, and the tiny ones, which one does not (usually) see by day and thus only knows, if at all, from books. If one is interested, that is. But what is merely seen in books remains unknown: it only becomes familiar with personal experience.

Starting with 1,000 watts

It was a butterfly collector, Hans Wimmer from the lower Bavarian Rot valley, long since deceased, who pointed out the diversity and beauty of our native moths to me when I was still a schoolboy with my final exams ahead of me. I marvelled at his collection without feeling the need to begin my own. I wanted instead to see the living moths and butterflies. Wimmer presented me with a massive 1,000-watt lightbulb, a mercury vapour lamp. With this I would go out on warm summer nights and illuminate the area from the small garden of my parents' house right out into the meadows that lay beyond our garden fence. The result was simultaneously overwhelming and bewildering.

In the dazzling light I saw any number of moths of every size and shape, although the largest and loveliest were much too restless under the glaring brightness for me to admire them properly. When they landed on a white bedsheet that was lit up by the lamp, it created such a massive contrast that my eyes soon became sore. Furthermore, I got a large quantity of insects in my face and hair and they crawled under my shirt, which was only briefly amusing and then very irritating. It was impossible to make any sense of the mass of insects that flew into and out of the light. After a few such illuminated evenings, I only paid attention to the particularly beautiful moths, ultimately giving the practice up when I started my university studies. This was partly because the summer semester kept me at university at the best time of year for observing moths and butterflies. Nevertheless, an enthusiasm for Lepidoptera had taken hold of me. From my third year of biology, I dedicated myself to the small and barely known species group of the aquatic moths (see pp. 15 et seq.).

How to successfully attract moths to light

While working on my dissertation on the small aquatic nymph moths, I learnt about a device that was used to catch insects for feeding small birds almost all year round at the Radolfzell Ornithological Station. The birds were used for research purposes at the associated Max Planck Institute. The principle is simple, but I would like to explain it briefly again: the light is provided by a fluorescent UV lamp (a 15-watt bulb) that attracts various nocturnal insects. There is a funnel or hopper under the light stick, and the insects fall down this and into a large sack, which is filled with egg cartons on nights when many insects are expected, in order to increase the internal surface area. The insects do not come to any harm. The next morning, they are gathered up to be fed to the birds and separated according to whether they are suitable or unsuitable for this purpose, or in order to determine their species and to count how many of each species had flown into the trap.

That was how I started. My first such device was built for me by Raimund Mascha, to whom I am bound by half a century of friendship. It worked perfectly from the first and I was therefore able to start with my own investigations in spring 1969. It rapidly became a very fruitful year – so fruitful that I soon wanted to investigate species diversity in the riparian woods and in other places using the same methods. These methods were subsequently adapted to a variety of situations. The work with the light traps quickly became routine. Despite its repetition in the same form throughout the year, over the years and over the decades, the thrill associated with the identification and admiration of the moths was never lost. Every night could bring something unexpected: for example, large hawk-moths, which I allowed to crawl onto my finger in order to examine them at close hand. I often showed people who were interested, including many children, how the eyed hawk-moth, *Smerinthus ocellata*, which holds its forewings over its body like a tent when at rest, responds to a light tap on the thorax by rapidly lifting and pulling them forwards. In doing so, it dramatically reveals its pink hindwings with their huge pair of 'eyes'. Even adults can be slightly shocked by this, and the sudden revelation of the eye markings clearly has a similar effect on birds, the main predators of this moth (see Photo 28). Like tiger moths and some other impressive moths, large hawk-moths will let themselves be placed, with careful handling, in the bushes, where they can spend the day protected until they wake up again towards late evening. There were special experiences – things that were completely new for me - that were very frequent at first, and although they became less frequent as my experience increased, they never entirely disappeared. In the early years it was the quantity and species diversity of moths and butterflies that impressed me the most. On the best nights, 500–1,000 moths flew into the light trap. There might be 150 or more different species per night. Sometimes I felt as if I was in the tropics; although I did not see such quantity or such diversity anywhere in Brazil, the tropics are especially species-rich, even though the individual species are mostly very rare.

It was not only moths that flew into the light but many other very interesting insects too. The cockchafers (also known as May bugs or doodlebugs) that swarm every three years and that were, at the time, so common, would sometimes clatter into the light like heavy hailstones. Luckily their main flight, which lasted two or three nights, took place so early (between late April and early May) that the beetle masses did not demolish the moths. This particular period was known as the 'April gap'. It separates the flight time of the true spring species, such as the common quaker, from the species of the early summer spectrum that are only seen from mid-May. But since the beetles came in their hundreds, the sack sometimes fell to the ground under their weight. The numbers on file for some of these cockchafer nights were far too low to give a realistic idea of their number, since many of them flew away without waiting until I came to count them the next morning. I shall return to the cockchafers, as their disappearance came at the beginning of the 1980s, just when the numbers of moths and butterflies started their persistent decline.

Change and continuity

Much changed around my home village during this time. Many meadows were converted into maize fields. One of the streams between the village and the forest along the River Inn was filled in and levelled, and all the surrounding woods with their enormous ashes, some well over a hundred years old, were felled. The half-kilometre-wide meadow next to this was turned into arable land that has since been planted almost continuously with maize. In early summer the wind would carry the stink of the pesticides that had just been sprayed on the fields into the gardens and often as far as the house, forcing us to keep the windows closed. The fruit trees around the edge of the village were for the most part removed by the state through the 'grubbing-up premium' scheme, as elsewhere. In many places the maize fields came right up to the garden fences of the village houses (see Photo 30). While in the 1960s, the village had lain idyllically between open fields that were used in multiple ways by the local farmers, now the village looked like an island in a sea of maize. In summer and early autumn, the two church towers projected over the tall maize plants, but from their harvest in October/November, through winter and into early May, the village was surrounded by bare soil.

The forests along the river changed far less and in a way that was barely discernible from the outside. The place where I began a second long-term investigation in 1973 lay more than 100 metres from the meadow, at the edge of a small settlement. The light from the moth trap shone directly into the woods. However, just as I began my research, the management of the woods as a coppice for the production of firewood, which had continued for hundreds of years, came to an end. Cheap diesel or heating oil had become far cheaper than wood and the owners gave up the arduous work of wood production. The low-growing trees that had been arranged in individual plots according to ownership thus began to grow together. For the first decade it was barely noticeable. It was only in my songbird counts that an increasing decline in the occurrence of most species became apparent. The age differences between the trees in the different plots gradually began to even out and the wood became more uniform. Initially, this structural depletion of the wood had less of an effect on the moths and butterflies than on the small birds, for whom the mosaic of differently aged trees was more critical. In the 1980s, I incorporated further areas in my investigations. One of them, right by a hydroelectric power plant on the River Inn, provided a wealth of data on abundance and seasonal swarming patterns for water-based insects (caddis fly, mayfly, stone-fly and non-biting midges that flew in great masses), but there were also comparable findings among the woodland species and moths and butterflies that lived in the dry grassland near the dam.

This area was not at all affected by agriculture. The same was true for a further investigation site, in a location that was at the time used by the German armed forces (*Bundeswehr*) within the managed forest. In 1981, I started comparative investigations in Munich in a completely unlit, 6,000-square-metre courtyard at the Nymphenburg Palace complex. This ended in 1984 when the Bavarian State Collection of Zoology (ZSM) moved into a new building in Munich-Obermenzing. There I could only begin investigations using light traps in 2002, but was able to continue them very regularly and in rapid succession until 2010. I carried out further investigations at various places in Munich. The different areas combined to create a kind of city cross-section, composed of parts of the city close to the centre and leading all the way to the countryside. In southeast Bavaria we were particularly interested in insects that were trapped in gardens, at the edge of the village, and in places close to nature – for example, along a stream valley: places that were not subject to the direct effects of agricultural use. Some gaps in the continuity of our records were unavoidable. The analysis of light trap results is very time-consuming. Weather conditions also had to be considered in order to achieve true comparisons. It is only possible to obtain meaningful in situ findings on frequency and population trends in nocturnal insects with a sufficiently high recording rate, that is, at intervals of two/three up to a maximum of five/six nights. Analysing a single summer catch can sometimes take hours. This enormous invest-ment of time is only worthwhile if it is possible to guarantee long-term continuity or if light traps can be set at different places on the same night. The ZSM therefore became the centre of our activities.

All praise to those who helped us with identification problems

Many people have helped us with the analysis of the light traps over the past 50 years. The trust I was able to place in the identifications provided by the ZSM from the start, especially from 1974, was very important. My colleagues there who were unfailingly helpful at the beginning were Dr Wolfgang Dierl and Josef Wolfsberger, both from the Lepidoptera Department, and the former Collections Director, Dr Walter Forster (all deceased). Thereafter, it was my colleagues and friends Dr Axel Hausmann and Dr Andreas Segerer who were particularly helpful, despite being extraordinarily occupied looking after one of the largest collections of butterflies and moths in the world. In the latter years, I must add Johann Brandstetter and Gerhard Karl, both amateurs but distinguished experts on noctuid moths and Microlepidoptera. Learning to distinguish between more than 1,000 butterfly and moth species in such a short space of time is not easy.

Since the species often vary, and sometimes very considerably, the reassurance of specialists is essential. Over the decades, a network was formed that helped to resolve uncertainties and correct misidentifications. New findings provided challenges. Some forms that were initially not recognized as separate species by specialists turned out, following modern molecular-genetic analysis, to have been a combination of two species. But this did not change the results – the spectrum of recorded species is too broad. We are not concerned here with the problems of differentiating such cryptic species or the separation of species twins. Our focus is to work out the general developments and trends and to determine what causes them. To this end, another preliminary observation must be made concerning the moths and butterflies themselves.

Butterfly and Moth Names

Butterflies are considered 'children of the sun'* and this poetic description suits most butterflies very well. Their flight begins in the morning, but only once it is warm enough. We can often see them warming themselves up with their wings spread out, facing the sun. By late afternoon they are already looking for a place to rest or hide for the night. They go to sleep, as a rule, earlier than us. Some butterflies and moths gather in 'sleep groups' for this purpose, in specific places only known to them, most probably marked with their own characteristic fragrance. They will then attach themselves in groups to the plant on which they have chosen to spend the night. Some blues behave like this, as well as the red and black six-spot burnets. In the tropics, there are many more species that form sleep groups.

Yet the butterflies that fly by day and the many moths that have adopted a diurnal rhythm, despite being moths, only represent a very small part of all Lepidoptera. In central Europe, butterflies make up only about 10 per cent of native butterflies and moths. The great majority of Lepidoptera are in fact 'children of the night'. This designation is justified not only by the numerical superiority of the moths in terms of species and populations, but also the diversity of families.

* In German, *Sonnenkind* refers to the happy, unburdened inner child.

Apart from the true diurnal butterflies (Rhopalocera, literally 'club-antennae'), only the skippers (Hesperiidae) fly by day; by night, there are hawk-moths (Sphingidae), tiger moths (Arctiidae), several other families of Macrolepidoptera, as well as noctuid moths (Noctuidae), geometer moths (Geometridae) and numerous families of micro moths (see Photo 29). These range in size from miniscule moths resembling dust motes to large hawk-moths with the body masses and flight capabilities of small birds. The fact that they are all referred to as 'moths' unfortunately detracts from their value. The pest species, such as the common clothes moth *Tineola bisselliella* or the Indian meal moth *Plodia interpunctella*, are small and unremarkable as moths, but their caterpillars cause irritating holes in unprotected fabric, or in stored food products. Using the same term 'moths', as we do in Britain, for all nocturnal lepidopterans (and butterflies for the rest) is so imprecise that it is almost akin to describing all mammals, including deer, bears, wolves and beavers, as either mice or rats.

The family names are generally more suitable, although with some exceptions. For example, the German name for the noctuid moths is 'owlet', although they have nothing in common with the owls of the bird world; and those moths that the Germans call 'spinners' do not all spin, although many caterpillars that do not belong to this group definitely do spin, and in fact very well. But we will let that be. The names are based on old impressions, opinions and prejudices, or magical interpretations, such as the German *Schmetterling*, which comes from the central German word *Schmetten* ('cream'), and the English 'butterfly', that probably stems from the similarity of the colour of the male brimstone to butter. We will have to accept the descriptions that have been passed down through the history of our language or work to ensure that they are gradually replaced by more suitable versions.

Some scientific names are no better, and several are actually misnomers. They express the vanities of specialists, contrary to what one might expect from science. It is absurd that species are named after people merely to flatter them. However useful they have been in their service to nature, their names need not be assigned to animals and plants. In my personal opinion this applies even to celebrities such as Charles Darwin. If one really cannot label a species using a generally understandable characteristic or offer a common quality for the description of the species, then using a geographical reference is still better than naming a butterfly or moth after a person. Anything else is arrogance, idiosyncrasy or flattery. The fact that these names exist and cannot be erased is due to the so-called 'principle of priority' that is supposed to guarantee the clarity and sustainability of scientific nomenclature: a system that was good in itself but that in practice has become unnecessarily burdensome.

The scientific name for a particular species is supposed to be determined for all time and in all languages, the first time that species is described. It should, moreover, be based on the original system for natural description developed by the Swede Carl von Linné (Linnaeus is the Latinized form) in the early eighteenth century and published in his magnum opus *Systema Naturae* for plants, animals and minerals from 1735. Volume 10 of 1758 is particularly significant for the animal kingdom. Since then, every species has been furnished with a double name consisting of a first name to describe the genus (which is capitalized) and a second name to specify the species (written all in lower case). In order to avoid confusion, each name should also include the name of the first person to describe the species and the date of that first description. The full name of our large white is therefore *Pieris brassicae* Linnaeus, 1758. *Pieris* is the genus of the whites, and *brassicae* is the species name, derived from the Latin word *brassica* = cabbage. Linné already listed it in the work on the animal kingdom referred to above, and so the scientific name of the large white is supplemented by 'Linnaeus, 1758' or abbreviated, 'L., 1758' or 'Linn., 1758', and is supposed to be unambiguous for all eternity. That would all be wonderful if contemporary botanists and zoologists had had access to the Internet and cultivated a regular exchange of information. Such correspondence would have made it possible to avoid the vast numbers of double and multiple (new) descriptions that were recorded, with the consequence that a great many species acquired multiple scientific names. Discovering the very first among all these names and finding the specimen that was the source of the description excites some specialists like the detection of historical crimes. This in turn has led, and still leads, to numerous amendments in the 'officially applicable' name. The problem is exacerbated by the ever-changing opinions of specialists arguing that a particular species should 'belong' to this or that genus or a newly created one.

Consequently, there are frequent name changes, always carried out to the best of the specialists' knowledge but, unfortunately, not on the basis of generally recognized, definitive methods. Accordingly, 'my' aquatic moth, for example, the brown china-mark, which used to belong to the genus *Nymphula*, which was still 'valid' in the 1970s and thus appeared as such in the publication of my doctoral thesis, has since changed to *Nausinoe* and then to *Elophila*, although Linné already knew the brown china-mark and named it in 1758. The very common and unmistakeable mother-of-pearl moth, whose caterpillars roll themselves into tubes made from nettle leaves, can be found under *Phalaena ruralis*, *Botys ruralis*, *Syllepta ruralis*, *Pleuroptya ruralis*, *Pleuroptya verticalis* and the currently 'valid' *Patania ruralis* – or perhaps not, given the confusion caused by such amendments, in stark contrast to the original principle of stability. Even such a universally known species as the peacock butterfly has been and still is affected, since it has been assigned a whole variety of genus descriptions.

Modern molecular genetics will do nothing to alleviate this misery; it is more likely to aggravate the chaos through the findings it has established because they are based on an arbitrary determination of 'genetic distances' for the justification of species differentiation. The hope that a consistent and clear system could be revealed using genetic lineages is simply an illusion. Everything is in flux when it comes to the process of speciation (the evolution of species). Even Charles Darwin, in his main work, *On the Origin of the Species* in 1859, especially in the sixteenth edition of 1872, the last that he worked on himself, fundamentally questions and denies the concept of clear, rigid and definable species. Species differentiation is a process that progresses with varying speeds and that places no hard borders between species, whether spatial-geographic or temporal-historic. It is thus purely a matter of opinion whether, for example, the butterflies or moths from an island or a remoter region of the continent should be categorized and named as a discrete species or 'only' as a subspecies, even when they are genetically or outwardly slightly distinct from another form that is clearly very similar and demonstrably closely genetically related.

Currently, the 'splitters' dominate the field and have done so for

1 *Red underwing* Catocala nupta *presents its hindwings with their deterrent red and black pattern. The forewings are camouflaged against the bark*

2 *'Shaking hands' with a death's head hawk-moth that has just emerged from its pupa*

3   The 'death's head' of the death's head hawk-moth: these are the actual back markings. Only an overactive imagination would recognize a human skull in these

4   Death's head hawk-moth caterpillar (Photo: Peter Denefleh)

5 Nymphula nymphaeata, *a male resting at the water's edge, photographed 50 years ago during the field research for my doctoral thesis*

6 *Lesser purple emperor on my hand*

7   *The purple emperor pushing its lemon-yellow proboscis into the pores of the rough skin on the flattened common toad; the red admiral is in the background*

8   *Purple emperor displaying the full glory of its blue sheen on a fern*

9 *The peacock butterfly: familiar, numerous and magnificent, yet still puzzling*

10 *Map butterflies of the summer form jostle for space on a flower during the population explosion of summer 2013*

11 and 12 *Map butterfly – spring form (above) and summer form (below)*

13   *Peacock butterfly caterpillars on a nettle plant*

14 *Painted lady – freshly hatched descendant of a butterfly that emigrated from North Africa in May*

15 *Small tortoiseshell – a misunderstood migrant butterfly*

16   *Silver Y moth – the most common migrant from the south*

17   *Red admiral – also a common migrant, although rarely seen in large numbers*

18   *Large white*

19   *A copulating pair of six-spot burnets* – Zygaena filipendulae

20   *A bird-cherry, eaten bare and covered in webs by ermine moths*

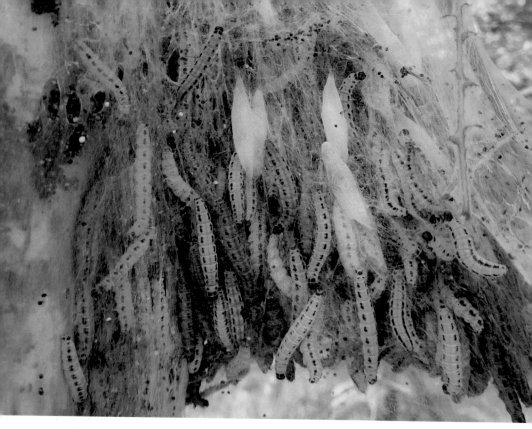

21 *Bird-cherry ermine moths – caterpillar masses that are starting to pupate*

22 *Ermine moths hatching en masse*

23   *Caterpillar of an elephant hawk-moth, in a threatening pose. When it pulls back its head, the eyespots on the front of the body and the twitching movements make it look like a tiny, threatening snake*

24 *Elephant hawk-moth: it is hard to believe that this dusky pink, streamlined hawk-moth is the result of the metamorphosis of a fat, snaky caterpillar (see Photo 23)*

25 *Brimstone.*
On a winter morning, covered in hoarfrost crystals

26 *Thawing (Photos above and below: Dr Eberhard Pfeuffer)*

27 *The privet hawk-moth,* Sphinx ligustri, *is one of the largest moths of Britain and central Europe and is still relatively common, thanks to the many privet hedges found in parks and gardens. Although a moth, it has the size and power of the smallest hummingbird*

28 *A pair of eyed hawk-moths mating*

29 *Burnished brass moth*, Diachrysia chrysitis (stenochrysis), *formerly known as* Phytometra chrysitis, *a typical noctuid moth* (Noctuidae)

30 *Field path: frequently a path that is used very intensively for agriculture will start right at the edge of the village*

31 *Box-tree moth: a questionable gain from imports from Southeast Asia. Their caterpillars defoliate box hedges,* Buxus sempervirens, *in gardens and parks*

32 *Marbled white,* Melanargia galathea: *one of the major losers among the butterflies of the open meadows and fields*

The riparian woods along the River Inn – a jungle of riparian woodland and water that has great species diversity

34 *The Zoological Collection in Munich, an inner-city oasis of species diversity. The library is under the meadow in the foreground*

35 *Wasteland with high species diversity in the local area – a habitat for many insects, especially wild bees, bumble bees and butterflies*

36 *Maize field at the beginning of June. The plants do not yet cover the ground, making them susceptible to heavy rain (and soil erosion). Later, when the plants are fully developed, other plants and insects have no chance of survival*

37   *Maize field edge without flowers: the end of biodiversity. Even inner-city carparks offer more for animals and plants*

38   *Sprayed to death all the way to the path edge. Overfertilized fields are treated intensively to remove the 'weeds' that are encouraged by the oversupply of nutrients*

39   *It is not only the edges of the forestry tracks that are mown at the end of June, but also the small yet flower-filled forest meadows. Suddenly the butterflies, moths, wild bees and bumble bees can find no more blooms: a maintenance measure in the state forests that is not just questionable, but that actually leads to additional costs*

40 *A dam that has been mown down to the ground. Nothing is left for the butterflies that the dam maintenance is supposed to serve. Bumble bees occur along the dams of the River Inn at less than a tenth of the levels of the 1970s, when no 'maintenance measures' were carried out*

41 *Silver-washed fritillary. Its caterpillars rely on the leaves of violets as their food plant, its butterflies on the blossoms that provide nectar. 'Roadside maintenance' in state-managed woodland destroys both. Raspberry bushes, a possible alternative, no longer grow on modern logging trails*

42   *A blue, a female Adonis blue, investigates a fingertip; a subtle yet memorable summer experience*

43   *Adonis blue,* Polyommatus bellargus, *male, a beauty that is disappearing from flower-rich fields and roadsides (Photo: Dr Hannes Petrischak)*

about 100 years, so that many species are becoming subdivided into two or more. Species numbers therefore grow, whenever 'new' species are recognized as such. However, they are not newly formed but have simply been discovered using a new method. A great many forms that were earlier recorded as subspecies or geographical races have now achieved the rank of species. The concept of species thus loses its 'traction', becomes less convincing, and the system becomes even more rigid than it was before. This would surely not please Darwin.

For good reasons, a countermovement was established at the beginning of the twentieth century, following a phase of extreme subdivision of species, which tried to condense this categorization and attempted to map out the nature of species using local and regional subspecies. The so-called 'lumpers' stressed geographic subspecies differentiations. Since even here there was some degree of arbitrariness, it should be noted that this method caused less confusion than the excessive splitting that had gone before. The currently predominant majority of systematists and taxonomists who are in favour of the division of species argue that this process gives rise to a greater number of species that are threatened and must be protected, since what was previously a local form is now a species. One may consider such justifications to be right and proper or dismiss them as being too personal and even tainted with ulterior motives, but either way, they do not alter the facts of nature.

Species are not so easy to pigeonhole, however desirable this might be. Their fundamental principle and the reason for their success is variation and variability.

The Decline of Moths and Butterflies

The village outskirts and the open fields

The assertion that there used to be many more butterflies and moths is confirmed by the results of my findings: the abundance of nocturnal Lepidoptera has declined by over 80 per cent since the 1970s. From 1969 to 1974 (except for 1970, for which there are no records), the average number of moths caught in the light traps per night between the beginning of June and the end of August was 242. Fluctuations in individual years varied from a maximum of 282 in 1972 to a minimum of 231 in summer 1974, and were thus quite limited. A mean average taken over five years therefore gives a good indication of frequency. From 1991 to 1995, it was 88 moths per night, in other words 64 per cent lower than at the start of the 1970s. Declines in Britain and North America have followed a similar pattern (see p. 220).

This quantity was then halved again, since the 37.5 moths that made up the average between 2013 and 2017 were only 15 per cent of the earlier frequency (see Diagram 1). The population explosions of ermine moths (see pp. 93 et seq.), which only took place in certain years and produced extremely high figures on certain nights, were not included in these calculations, nor were the figures for the box-tree moth (or box-worm moth), *Cydalima perspectalis*, that has been introduced to central Europe

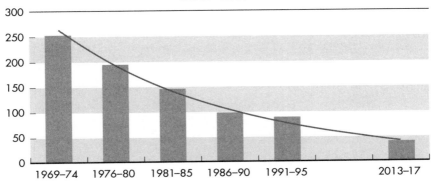

Diagram 1: Decline in Lepidoptera abundance at the edge of the village 1969–2017

*Exponential decrease of over 80% of nocturnal moths at the outskirts of villages in southeast Bavaria over the past 50 years
(average catches in the months of June, July and August, cf. p. 152)*

in recent years (see Photo 31). Their high figures would alter the results for the area around the outskirts of towns and villages significantly, and disguise the trend. The trend itself is best represented graphically, and I have therefore included a few graphs in this chapter. This finding is confirmed and corroborated by the decline of the remaining insects that do not belong to the Lepidoptera but that also fly into the light traps – see Diagram 2. This diagram does not include the swarms of cockchafers that ended towards the start of the 1980s. These used to take place very regularly every three years between the end of April and the start of May. They will be brought into the discussion later.

The decline in the other insects has in fact been much sharper than that of the moths. The number that flew into our light traps during recent years was only 4 per cent of that of 50 years ago. The finding confirms what car drivers have long known, since their windscreens now only occasionally need to be cleaned to remove squashed insects. This has nothing to do with the enhanced streamlining of modern vehicles and everything to do with decreases in insect populations. If insect population levels were unchanged, then the steep windshields of buses and lorries as well as the headlights of most makes of passenger car ought to be affected as before. The fact that the decline of moths by 85 per cent is slightly less than for the remaining insects (96 per cent) has a simple

Diagram 2: Decrease in abundance of other insects at the edge of the village

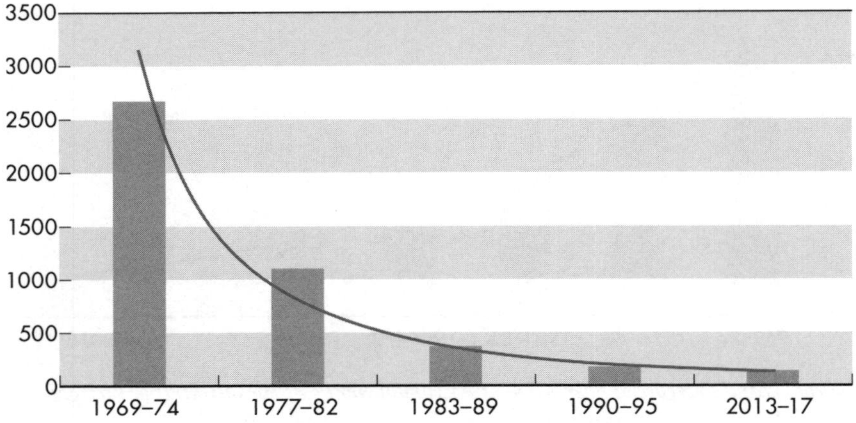

Even stronger exponential decrease of insects other than Lepidoptera at 96 per cent as measured using light traps since 1969
(average numbers measured as for Diagram 1)

reason. Moths, in particular the numerous noctuid moths, are good at flying. They fly into the light traps from significantly further away than most small insects. We can assume that moths will typically fly three times as far to reach the light traps than most of the small insects. Axel Hausmann discovered this with his doctoral research. With a larger catchment area, there will inevitably be a relatively larger quantity of insects.

However, both sets of findings clearly coincide to show a trend: abundance has dropped exponentially. This reinforces the impression gained by those nature lovers who are old enough to remember the 1960s and 1970s. It is true that there once were many more moths and other insects than there are today. Admittedly, we still need to deal with the objection that our memories relate to the diurnal species, especially butterflies. Unfortunately, it is not possible to record these as perfectly and transparently as the nocturnal species with their attraction to light. As a physical method, the use of light traps works independently of their users and is equally reliable no matter who uses them. Errors in findings can influence the results when it comes to a few species (see below) but not the assessment of quantity.

Counting butterflies requires a standardized procedure. The best method is the so-called 'transect survey'. However, in analysing the findings obtained in this manner, much more depends on being able to guarantee comparability. Time of day, weather and land-use, as well as weather trends, cause much greater variations in the findings by day than with the numbers flying to the light traps at night. Add to this the fact that the most common and well-known butterfly species are either partly or wholly migratory and that their observable frequency at a given place may therefore have little or nothing to do with the place itself. Those that are more closely bound to a particular location, such as the blues, the marbled white, *Melanargia galathea*, and some other typical meadow butterflies, are more suited to long-term comparisons. The peacock butterfly, small tortoiseshell, red admiral and painted lady and also the cabbage whites should, however, be considered separately. Diagram 3 thus takes into consideration only the native meadow butterfly species. This includes dozens of different species, more than 120 in total.

The decrease in meadow butterflies is therefore of about the same order of magnitude as for the moths. In years with high precipitation in May and June, where these months are especially unfavourable for meadow butterflies, the numbers fall further and amount to barely 10 per cent of earlier records. But even a decline of 90 per cent belies the extent of the actual loss. In 'high-performance' cultivated pasture that is mown

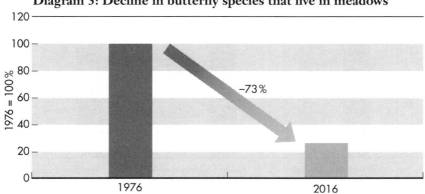

Diagram 3: Decline in butterfly species that live in meadows

Compared to 1976, the abundance of 'meadow species' of butterflies had decreased by 73 per cent in 2016

four or five times a year, no butterflies can survive any more. At best, a few will fly over the area from time to time, but this does not really 'count'.

In determining such trends, we should therefore not really include all the moths, butterflies and other insects flying into the light traps at night that come from gardens rather than the actual field that is being investigated. The recording method using light attraction at the village outskirts probably produces results that are far too favourable with respect to the fields. Actual losses would be even higher than 85 to 96 per cent, the final values at the time, if the numbers from the garden populations were excluded. The consequences for field birds are discussed on page 225. What needs to be clarified now is whether these extremely high declines relate to special conditions in the fields and meadows of southeast Bavaria or to a more general phenomenon (see Photo 32).

In the search for causes it is extremely important to differentiate between general effects, such as climate change, and special ones, which are independent of this. Clarity on this key question is provided from the findings that have been produced using the same methods in forests, in the city and within the villages, at locations that are sufficiently distant from the open fields. Diagram 4 provides significant evidence that condi-

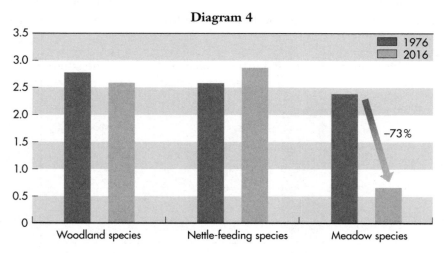

Diagram 4

Changes to frequency (abundance) of forest species and nettle-feeding species (whose caterpillars feed on nettles) as compared to meadow species, in each case where 1976 figures are taken to be 100%

tions are very different for butterflies from the forest and for those whose caterpillars live on nettles compared to the meadow species. Compared to the latter, forest species have remained unchanged over the last 40 years, allowing for normal fluctuations. The nettle-feeding butterflies actually showed a slight increase, while meadow species lost 73 per cent of their number. 'Meadow', as used here, indicates open country without tree cover and pasture in the normal sense of that word but not arable land. Dams or embankments make up one part of the comparative transects for determining butterfly abundance; the other is permanent pasture used for agriculture. Meadow species also live along roadside verges and embankments, as well as on wasteland and along cattle-tracks. We can therefore characterize them as the butterflies of the open country. As such, they only partly correspond to the nocturnal moths that fly into lights on the outskirts of conurbations. The expression 'forest species' is far more appropriate, and 'nettle-feeding' is unambiguous. These findings thus lead us to another sort of biotope, the riparian woods.

Findings in the riparian woods

As I have already mentioned above, in 1973 I began a second series of long-term investigations right at the edge of the riparian woods, at the lower River Inn. These areas were not affected by what happened in the fields, since they were far enough removed. But just as at the edges of towns and villages, there were normal garden plots here, from which moths flew into the light traps in a similar manner and with a comparable range of species. This was easy to deduce from the species spectrum. Nonetheless, the vast majority of moths doubtless came from the forest itself, and what we collected there on warm, damp, dark summer nights was most impressive. With well over 100 species – at peak times more than 150 – they exceeded the best catches from the village outskirts by almost a half (see Photo 33).

As can be seen from Diagram 5, a decrease in butterflies can also be confirmed for the riparian woods, although it is not as extreme as on the outskirts of the villages. If one takes the average for 1973–5 as the reference basis (that is, 100 per cent), then the values for 1991–5 are 40 per cent lower. However, given that there was too little data for the main flying time in 1973, at the start of the investigation, the implications of

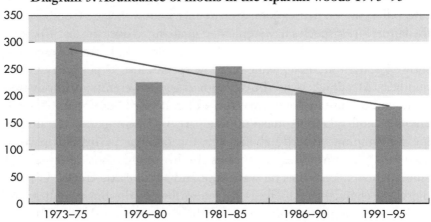

Diagram 5: Abundance of moths in the riparian woods 1973–95

Decrease in abundance also shown in the moths of the riparian woods (data procurement as for Diagram 1; cf. text)

the averages for the subsequent five-year period are not as clear. The actual decrease in frequency might perhaps have been a third. Still, the riparian woods were preserved and not cultivated: they are still there.

An explanatory note should be inserted here. The inferences that may be drawn from such statistical findings depend considerably on the selection of time periods for comparison. If the numbers caught in light traps, averaged out over the months of June, July and August, were reproduced for individual years as a series, clear decreases could be confirmed by means of an appropriate selection of years. This would be true, for example, for the years 1974–81, 1982–90 and 1991–5. But then there was a marked increase from 1977 to 1983. The fluctuations from year to year are always high: the deviations (statistically classified) are greater than the mean. Grouping the years together does temper these variations to some extent, but then no statistically secure trend may be evident over the total time span. Five years, even a whole decade is actually not long enough to establish ongoing trends and remove fluctuations, if the frequencies fluctuate very strongly.

It is thus quite possible that the Lepidoptera frequency in the riparian woods has not actually changed substantially and that the findings for the many moths coincide with the findings for the far fewer species of butterflies in Diagram 4. However, before we consider the possible reasons, it would be worth taking a closer look at the findings from Munich, in order to avoid premature speculation that could take us in the wrong direction.

The findings from Munich

Between 1981 and 1984, I had the opportunity to carry out investigations using light traps in Munich under very special conditions, namely in the inner courtyard of the North Wing of Nymphenburg Palace, which was where the Bavarian State Collection of Zoology (ZSM) was housed at the time. The approximately 6,000-square-metre inner courtyard had the advantage that it was not exposed to any artificial light at night. The palace buildings were so high that they effectively shielded the space: indeed, they were so high that the highest branches of the oak tree that grew in the centre of the courtyard barely reached over the roofs, despite their splendid height. The courtyard was unmanaged and, other than some small, irregular mowing that did not take place all year round, was left to itself. The window of my office looked out onto this courtyard. Between 1981 and 1985, when the ZSM moved to its new building in Obermenzing, I used technically identical methods to investigate nocturnal insects using UV light. We had to wait until 2002 before the investigations could be continued. In the meantime, I supplemented them using various different locations in Munich, which, when taken together, produced an ecological cross-section of the city.

I began the second continuous record on the area of the ZSM in 2002. It continued until 2010. The building of the ZSM is largely set underground. Grass grows over the specimen storage buildings and over the main area of the surrounding open space. At the eastern edge there is a small wood. It is the last remaining stand of the old coppiced forest, used to provide bark for making leather, dating from a time when Munich did not extend so far to the west and the palace grounds of Nymphenburg Palace formed the perimeter of the city. An overgrown hill, shaped like an oversized flowerpot, towers over the Collection premises, and in front of this there is a pond that is also quite overgrown. Seen from above, the otherwise unmanaged, almost rectangular plot of land gives the impression of an island in the middle of a typical metropolitan residential area. Gardens surround almost every house. A suburban railway line borders the plot to the east. A further one passes a couple of houses further to the west. To the south, there is a large and busy road leading out of town to the west and to the motorway.

And yet, the premises of the ZSM do not quite form an island, since the gardens of the surrounding area also contain much greenery and the suburban railway lines offer further stretches of habitable land for insects and other small animals, since they go past the 180-hectare Nymphenburg Park. What can be seen of the ZSM above ground gives a futuristic impression, with its glass and titanium-clad peaks. The offices are built around two atriums, set over two subterranean floors. The specimen collections are all completely underground. These special circumstances do not disturb the birds that live on the site or regularly visit it. The site is left to itself, without being subject to any rigorous maintenance system, and in some years there is no maintenance at all. It was acquired in 1985 and for exactly a quarter of a century I had the privilege of being able to work there, or perhaps I should say, being domiciled there. I joined in 1974, when it was still housed in the North Wing of the Nymphenburg Palace.

The globally important collections kept there offer one inexhaustible reservoir of interesting things; another is the natural space that is developing on the site above. Both provided inspiration on an almost daily basis. The diversity of this living world could hardly have been united in a better manner: downstairs the collections, upstairs real life. Almost everything of importance that has been published on the widest range of animal species could be discovered and studied in the library, one of the most significant zoological libraries in the world (see Photo 34). In the time before the Internet and the lightning-fast exchange of information around the world by email, the collections and the specialist literature housed there were the most important foundations for zoological work. And as I gave lectures at the Technical University of Munich on landscape and urban ecology, it was convenient to connect the resources of the ZSM site with my research on urban nature.

In 2002 I could finally return to my investigations using light traps with nocturnal insects that I had started so successfully at our previous accommodation in the inner courtyard of the old Nymphenburg Palace, and I continued them until 2010. What they produced was surprising in

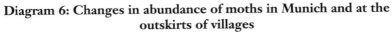

Diagram 6: Changes in abundance of moths in Munich and at the outskirts of villages

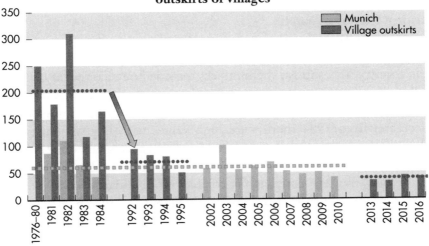

While the abundance of moths in Munich has not changed (in a statistically significant manner) since 1981, it declined steeply at the village outskirts and currently lies below that for Munich (production of annual average values as for Diagram 1)

more ways than one, and very instructive. The most important finding was provided by the comparison with moth abundance at the village outskirts (see Diagram 6). While there had been significantly fewer moths in Munich between 1981 and 1984 than outside the city, on the edge of the village, the numbers recorded there for 1992–5 were, to a great extent, similar to inner-city conditions. Thereafter, the numbers for the village outskirts sank again. With a summer average of 37.5, they are currently more than 25 per cent lower than the level of, on average, 51 specimens per night in Munich. A city with more than 1.5 million inhabitants, Munich is thus no less favourable as a habitat for moths than the open land in southeast Bavaria. The fact that the weather had a particularly positive effect on them there, leading to the exceptional moth years of 1982 and 2003, also emerges from the diagram. Even in the isolated inner courtyard of the Nymphenburg Palace, the year 1982 provides a significant contrast, just like the 'super summer' of 2003. I shall discuss this and what we can deduce from these results about the effect of heat on moths in due course. Before that, it is important to stress that there was no particular trend in the abundance of nocturnal moths in Munich between 1981 and 2010. This is also evident from Diagram 6.

The abundance fluctuations for individual years in the city remained within the range of normal variations: in fact, they varied less than in the country. In contrast, the numbers recorded outside the city, at the village outskirts, clearly decreased. The mean averages (the straight rows of dots shown in Diagram 6) indicate the three stages of the decrease: from 1984 there were high values with large fluctuations, then much lower values, but with smaller fluctuations, and finally, since 2013, there has been a frequency level that is less than that for Munich or at best roughly corresponds to it. The abundance of the other insect groups, which is also significantly higher in Munich, also points towards the further decrease in the countryside. This will surely vary greatly from place to place, yet it will not contradict the fundamental finding, which is that a massive disappearance of insects has taken place in the countryside. This emerges consistently from many individual investigations that cannot be reproduced here, but which are available on the Internet.

But now let us turn to the special years of 1982 and 2003. The first, 1982, can be distinguished from the years that preceded and followed it (see Diagram 6) both at the village outskirts and in Munich, as well as in the quantities trapped in the riparian woods. It was a special year for moths. But, as was found following the 'super summer' of 2003 on the site of the ZSM, after 1983 the numbers recorded for all three very different areas again declined steeply and returned more or less to the previous level. 'Singularities' such as these do not appear to bring about permanent consequences for moth populations.

A comparison between 2003 and the years before and after is set out in Diagram 7. The major peak is seen in June. In 2003 there were already so many moths in that month that the figure was more than three times that for the same month in the preceding and subsequent years. This led to a far greater volume of moths later in the year than in other, normal years. The extraordinarily favourable weather conditions of June 2003 were the key factor behind these numbers. It was 'hotter than ever', as reported by some media sources. The almost Mediterranean weather favoured the moths of the first generation. Consequently, a particularly

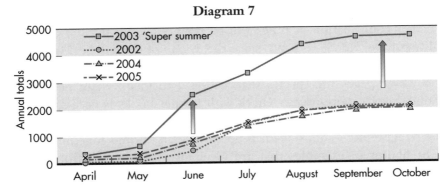

The cumulative development of the annual numbers of moths recorded using light traps in Munich (at the Zoological Collection) for the years 2002–5. The curve for the extraordinarily warm summer of 2003 is very distinct, in particular in the high increase of the June values

good second generation came about. However, when the weather in the following years returned to its normal changeability, the effect was lost. The more exotic species that were not seen prior to that year no longer appeared, and none of them, not even those from Southern Europe, established themselves in permanent populations.

But what if we did not just consider the quantities of moths, or their frequency, but the underlying species diversity? It could be that there are many species of moths in the city, like the sparrows, tits, pigeons and crows, that are widely dispersed and numerous, and that, thanks to their adaptability, are able to live in particular abundance in the unnatural environment of the city. Moth *numbers* alone reveal nothing about whether the statistics are species-rich or species-poor. The level of species diversity that exists in the village outskirts in the countryside and in the riparian woods should also be clarified, together with any possible relationship to the changes in the numbers of moths.

The decline in species diversity

Early in the morning I am always filled with tense expectation to see what the previous night's light traps will hold. In summer there is also the challenge of dealing with the swirling mass of insects. No moth may fly away without first having been identified and registered.

Identification requires proficiency. Even after decades of practice, some further confirmation or other will be required following the count. I always have 'Koch'' to hand, since this is one of the handiest yet most comprehensive identification guides available today. I have long known on which pages the species that are hard to identify can be found. Sometimes a more precise check with further books is necessary. My current favourite is *Die Nachtfalter Deutschlands* [*The Moths of Germany*] by Axel Steiner et al., since it is up-to-date and brings together excellent photos of examples that are ideal for identification from moth collections and images of living specimens, shown in their customary positions. Naturally, one learns from year to year and becomes more confident and more efficient, but one should not be too self-assured. There are always difficult cases and always new ones. This is part of the attraction.

Over the years, impressive species lists took shape. For the riparian woods in southeast Bavaria I recorded more than 650 different species of moths in the 23 years between 1973 and 1995. In the village outskirts I counted about 500 over 25 years, which was not quite as good but showed a wide diversity of species at the time (see Photo 35). This has long since changed. The species numbers that have been added up over the years might seem impressive but can also be misleading. This is because what appears over the years to be a new species need not come from the location where it was found and its immediate surroundings. Moths that are attracted by our lights might have been carried there from elsewhere by the wind. The fact that they can fly is a prerequisite for being caught using this method, but it also creates uncertainty regarding their provenance. Many species that are 'verified' do not necessarily belong to the local fauna.

The majority of those that do fly in from elsewhere – or are borne in on the wind – are rare, naturally, since we are dealing with host specifics whose caterpillars only occur on certain plants that are themselves rare, or because they happen to have flown in from further afield. If one sorts out the records of the spectrum produced according to the abundance of species, then after a few years a characteristic sequence usually becomes apparent. There are a few very

---

* Manfred Koch, 'Wir bestimmen Schmetterlinge' (Neumann-Neudamm, 1984).

common species that occur every year and which, when the figures are collated over about a decade, produce correspondingly high total figures. Significantly more species fall into the group of medium frequency. They range from continuously occurring to the rare, with just a few individuals. The largest proportion is made up of the many other species of which only singletons occur. In other words, there are many from just a few species, and then a few each from a great many species. Just how these are evaluated with respect to the local lepidopteran world is a matter reserved for specialists. One cannot determine from the outset below what (slight) level of occurrence the species are no longer considered part of the 'fauna'. In this situation it is only possible to establish a final value for all the species in an area mathematically: to produce a figure that will become more exact over the course of several years. However, since their living conditions very rarely stay the same over (very) long time periods, this ceiling value itself will also change.

The problem is therefore how to solve or avoid these difficulties in order to produce meaningful results in respect of the various species. There are two suitable approaches. One of them assumes that, for most investigations into moths and butterflies and other insects in the open, it is unrealistic to attempt to record 'all the species'. If one considers the accumulation of the spectrum of species over the years and its proximation to a ceiling value, then it is simply more sensible to be satisfied with a 90 per cent recording. It is then relatively easy to calculate how long it will take until this 90 per cent is attained in the study area. For the species-rich riparian woods, for example, this would take around seven years. The considerably less species-rich, enclosed inner courtyard of the Nymphenburg Palace, by contrast, only required four years for the 90 per cent ceiling to be attained. With larger areas that have a mosaic of contrasting biotopes, one should logically allow at least a decade until the recording level of 90 per cent has been reached.

On the one hand, these are clear time requirements; on the other, they can be quite unrealistic in practice. For example, where investigations are supposed to clarify what will be changed by the interference of construction work, prior to its commencement. The long time period necessary for this is usually not available. Consequently, short-term investigations may deliver findings that are not 'stable' enough to be used as a basis for a comparison with what will develop after the work has taken place.

Naturally, this statement applies in principle to all changes, not just those caused by road and other construction projects. It is important for me to explain and emphasize the underlying ecological principle here, since it could be argued that the results from the light traps are not representative because nature is too variable. But all three focal areas where light traps were used clearly fulfil the criteria of (at least) 90 per cent documentation, since research was carried out there with respect to the local species diversity for five or seven years or (much) longer.

The second option for comparison is simpler and more suited to shorter time spans. The concrete species numbers in the light traps per night are taken as a basis, during the main flight period of June to August or another comparable period, just as for the determination of abundance and its changes. It will be easiest – and most meaningful – to compare the findings if the investigations are carried out simultaneously. The average values will be heightened due to the particularly high species counts following good nights. This effect is diminished if the average highest values are taken as a basis and those that were low due to unfavourable weather (cold, bright moon, strong wind) are ignored. From a statistical perspective, I would like to point out that these calculations do not correspond to any so-called 'normal distribution' since the relevant flight conditions each night have a strong effect on the quantity as well as the species count of the moths that are recorded. The statistical fluctuations are thus always significantly higher than with a 'normal distribution'. However, the comparability is not affected, since all the diagrams were created in this manner.

It is apparent from Diagram 8 concerning species diversity and its changes that the average highest values from the village outskirts showed no particular trends until the early 1980s, but thereafter decreased sharply. Compared to the values from the 1970s, they were halved. Today there are only about half as many species of moth as there were in the 1970s. This finding is consistent with the much sharper decline in abundance of 80–85 per cent. The extremely sharp decrease in moths is therefore not based predominantly on the fact that a few less frequent, though common, species have become rarer, as one might assume.

Diagram 8: Lepidopteran species diversity

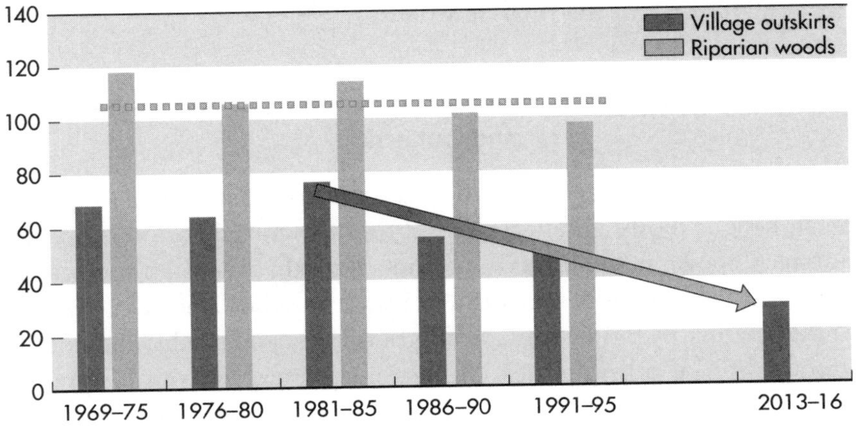

Change to average numbers of species counted in light traps in the riparian woods (no decreasing trend) and at the village outskirts (significant decrease since the 1980s)

Many of the rarer moths have also disappeared and are still missing. This relationship is logical. First, the abundance decreases. Only then do the species start to disappear from the spectrum one by one. Those that are already rare are more likely to be affected than those that are still common. The range of species is uniformly compressed (= loss of frequency) and 'trimmed around the edges' (= loss of species count). The cause of the decline must have started in the fields in the 1980s, since there is no change in species diversity with the moths and butterflies of the riparian woods until the end of the investigations there in 1995. The fact that the moth quantities had also markedly fallen in the riparian woods of the lower River Inn (see Diagram 5) did not (yet?) have an appreciable effect on the species range at the time.

The days when hundreds of moths would fly into the light traps, representing 150 or more different species, are therefore over. The reduction in species diversity in the fields was especially notable, much less those that lived in the riparian woods and least of all the species range in the city. It might seem that the moths and butterflies that occur in commercially managed woodland have hardly been considered. Numbers were recorded over many years in the late 1980s and early 1990s at a base belonging to the German armed forces. To go into these in more detail would, however, be beyond the scope of this book, especially since they would only serve to underline what has already been established for the riparian woods.

Warm summers and what they mean for the moths and butterflies

With that, I would like to return to the warm summers and what they mean for our moths and butterflies. From the consideration of species numbers above, it should have become clear that the high increase in numbers in the super summer of 2003 was caused by a positive reaction to the weather by the already existing butterflies and moths, and not the immigration of other species. Moreover, it emerged from Diagram 7 that the question of whether it will be a (very) good year for moths and butterflies or not depends on the conditions in June. If it rains a great deal and if May and June are generally too cold, the remaining three months of 'summer' (that is, the summer according to the calendar) will be unable to compensate for this negative impact. The old saying 'one swallow does not a summer make'* gains a new, broader meaning in this context: even a good summer will not bring about a change. There would have to be several good summers, in fact many, and they would have to be consecutive.

Furthermore, the reaction of the moths and butterflies also reflects the fact that warmth is good for them. It poses no intrinsic threat to the insects. This is especially true for the nocturnal species, which after all make up the vast majority of our Lepidoptera, even though they still remain largely unknown to the general public. Their reaction to warmth is like our own. If that were not the case, then surely millions upon millions of people would not travel south every summer to spend their vacations in the even greater heat there. In many Mediterranean locations people will come across moths and butterflies in quantities that remind them of 'earlier' times when they were still abundant closer to home. There are considerably more butterflies in the warmth of the Mediterranean because summers in that part of the world have

---

* Aristotle (384BCE–322BCE): 'One swallow does not a summer make, nor one fine day; similarly, one day or a brief time of happiness does not make a person entirely happy.'

Diagram 9: Temperature changes in southern Bavaria (°C)

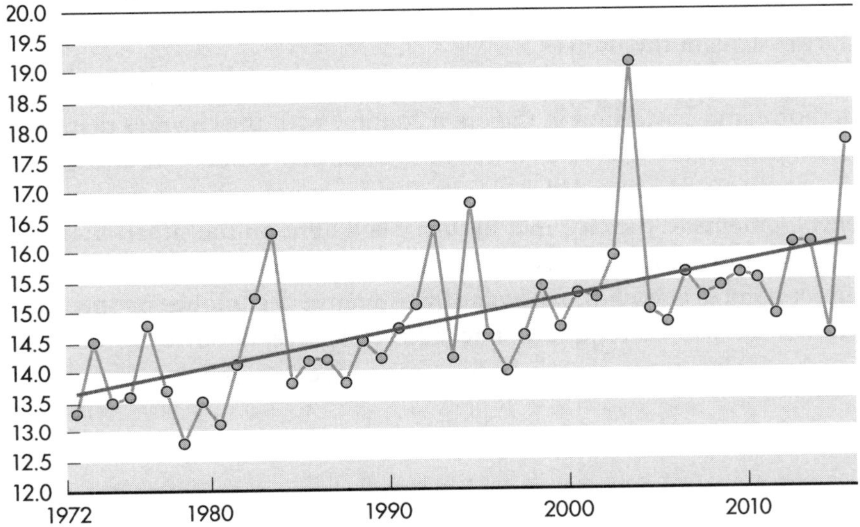

Increase in average summer temperatures in southern Bavaria since 1972
Source: *German Meteorological Station, Hohenpeissenberg*

always been not just reliably early, but also hot. The butterfly world has adapted to this. We did not have such temperatures in central and northern Europe when butterflies were far more common than they are today. The weather was changeable and not nearly as favourable as it was around the Mediterranean.

The fact that butterflies were more common 50-plus years ago than in the best recent butterfly years, even though it was colder in those days than it is today, looks like a contradiction that merits our attention. It is without doubt not due to global warming, since this ought to have led to more butterflies than in earlier, colder times. Diagram 9 under-lines this conclusion: despite increasing average summer temperatures since the start of the 1970s, the frequency of moths and butterflies has instead decreased (compare with Diagram 1). They can clearly only bring about a population increase when there is a very hot (early) summer. The increase in temperature of several tenths of a degree every decade, contrary to what one might expect, does not appear to have had a beneficial effect. This fact seems even stranger if we consider the circumstances in the city more closely and compare them with the (surrounding) countryside. The results are very revealing. Even the slighter decreases in butterfly populations in the (riparian) woods will start to explain themselves.

This finding again connects the search for the causes of the decline in moths and butterflies in the open country with the changes that have taken place there over the course of the last fifty years. The changes in the countryside must differ significantly from the circumstances and developments in the city and, in turn, shed light on the situation in the forests and woodlands. Let us therefore consider town and country in direct comparison. No particular investigative techniques or specialist knowledge will be required. The differences are so obvious, yet since we are used to them we barely even notice they are there.

The Metropolis: The End of Nature or Salvation for Species Diversity?

Stereotypes feed the widely held preconception that the countryside equals 'nature' and is therefore 'good' and that the city is manmade and, as such, 'bad'. In 1965, the influential sociologist Alexander Mitscherlich published a book with the title *Die Unwirtlichkeit unserer Städte: Anstiftung zum Unfrieden* [*The Inhospitality of Our Cities: A Deliberate Provocation*]. It is true that he used the book to pillory the living conditions of the 'proletariat', but the emerging German nature and environmental conservation movement of the 1970s took up the concept of 'inhospitality' in order to campaign against the cities and their rampant growth. For them, the big city spelled the end of nature. The cities were devouring the good land and for many this could not be ignored.

Since then, this attitude has characterized the position of nature conservation organizations in Germany. In this, they summarize the ideas of the German Romantic movement of the late eighteenth and early nineteenth centuries, in the words of the hiking song: 'We leave behind the grey city walls and roam through woods and fields, whoever stays can rot.' The countryside, idealized in those days as the counterpart to the city, inspired the environmental movement in its choice of campaign colour: a vibrant green, rather than the dark green of the forest, the hunter and managed woodland. The Green Party used this environmental green to symbolize their political position. For many years, that party has been fighting alongside nature conservation organizations against urban growth. They hope drastically to reduce 'urban sprawl'. In order to meet this undoubtedly real need, they favour so-called 'densification', that is, the higher concentration of building and development in existing settled areas. Open areas in the cities should be built up to avoid the loss of the surrounding area and further encroachment on the natural environment. This Green ideology is a simple extrapolation of Mitscherlich's 'inhospitality of the cities': the good land must not be further devoured by the evil cities.

And yet, the occurrence and abundance of wild animals and plants tell a quite different story. The cities are species-rich, and appear to be richer, the larger they are. In cities of much more than a million inhabitants, such as New York, London and Berlin, there is such a rich array of animal and plant life that they would easily qualify as nature reserves or even exceed them in terms of species numbers. Indeed, the wildlife of such cities has recently become a subject of television documentaries, and surveys of the green spaces of such metropolises, for example the gardens of Buckingham Palace in London, are carried out in detail. They also provide eminently important retreats for species that are threatened with extinction. In our context, this is evident from the fact that there are currently not only more moths and butterflies in cities than in the countryside, but that they represent a wider range of species.

I will stay with the rigid generalization 'in the country' for now, but I will shortly explain in far greater detail what this is supposed to mean. Birds in the cities twitter and sing in praise of their city and, at night, inaudible to human ears, the bats do the same with ultrasound as they hunt after insects. For botany experts, the rich diversity of wild flowering plants in cities has long been established. The urban area of Nuremberg, where detailed investigations into flora have taken place over several decades, contains twice the number of plant species as an area of the same size in the surrounding countryside. These investigations did not even take into account species planted in gardens. If they had, they would have at least doubled the inner-city species diversity of flowers.

Evidently, botanical diversity in the urban habitat does not correspond to the preconceptions of some botanists and zoologists regarding the composition of the flora. But we are not dealing with habitats that arise naturally, without the intervention of humans. City nature is something new, and yet the diversity is the response of the plants and animals themselves, since everything that grows and flies, walks or crawls, has come 'by itself'. As unbelievable as it may sound, this statement is so true that the fauna and flora in the cities are more natural than those of the commercial or managed forests, since nobody replants the wild plants or summons the city animals. The new cast of characters demonstrates just how flexibly 'nature' reacts to changed conditions. The fact that this unexpectedness is displeasing to botanists who are predisposed towards potential natural vegetation is a result of their expectations about what nature should and should not be. The fraction of zoologists who are concerned with the fauna of a wide area or a large biotope tend to be more flexible, since the size of the area makes it more difficult to be prescriptive about where animals are supposed to live. Nevertheless, there are strong divisions here too, caused by those who reject the new and support the eradication of invasive 'alien' species. But enough of these sweeping remarks on city versus country. They are enough to frame the question: 'what are the ecological differences?'

The advantage of structure

In the first place there is the structured character of the city. Aerial photographs or, even better, satellite images, such of those on Google Earth, can show at a glance what this means. A town is organized or subdivided into buildings, traffic routes, open spaces, parks, copses and often water bodies. The larger it gets, the more structure the city has. This can be measured directly by the length of the boundaries between the structures. For the first approximation, the traffic network is enough, combined with the number of individual buildings. In the next one, the open spaces are included. This ranges from land that has not been built on and is not otherwise used, to smaller pieces of greenery, gardens, parks and groups of trees and on to copses, woods and water bodies. Ecological science can use this to measure structural diversity and express this in figures. It is one of the central findings of ecology that the diversity of species directly depends on structural diversity.

Uniform, relatively unstructured biotopes are species-poor; diverse and many-structured ones are species-rich. According to this principle, it would be logical to encourage the replacement of monocultures in commercial woodland, such as pure spruce stands of the same age, with more species-rich and more structured forests.

The situation in the countryside is quite different. Wherever there is a space that has not been subdivided geologically by hills, cliffs, gorges or rivers, closer inspection will often reveal that the land is quite uniform, even monotonous.

This was not always the case. Structural subdivision only started to diminish with the introduction of land consolidation. This led to the creation of ever-larger, agriculturally uniform areas. Maize or rapeseed oil fields can stretch all the way to the horizon, and even beet and wheat fields are no less uniform. Comparing similar surface areas, villages, small towns and, above all, large cities have a far higher level of structural diversity than agricultural areas. The same can be said of large-scale forested areas. Most of them are not woods in the natural sense of the word, but commercially managed forest monocultures. However, to emphasize my point once again, it is this structure that is the basic prerequisite for biodiversity. Where nothing but maize or rapeseed grows over many square kilometres, or where spruce stands make up a whole managed woodland, biodiversity cannot prevail (see Photos 36 and 37). Only very few specialist species can manage under such monocultural conditions and they are almost always pests, because they have adapted themselves biologically to the plant species that are available on the same scale. Their nutritional basis is so plentifully available that they can reproduce in their masses.

Monocultures produce pests

This leads to my second point: the homogeneity of the 'offering' favours the mass proliferation of the user. Commercial losses are practically pre-programmed when monocultures are used. They can only be held in check inadequately, at best in the short term, by applying appropriate, usually highly poisonous repellents. In the highly structured landscape of the city, similar cases of damage virtually never occur. Even when individual species of crop plants or the trees in a city park are infested by pests, it is incomparably easier to deal with such an infestation here than in the massive areas made up by managed woodlands or fields. Often, intervention is not even necessary and, if anything, counterproductive. This was shown by the case of the bird-cherry ermine moths. The use of poison as a means of controlling undesired population explosions and for defence against damage can be dispensed with.

The risk of pests infesting and destroying crops in the field is hugely magnified by the genetic uniformity of the variety used. Under these conditions, the fight against pests effectively selects pesticide-resistant variants much more quickly and more comprehensively than in populations with high genetic diversity. We shall return to this. In any event, what is considered a basic market law in economics applies here too, namely, that supply and demand are closely dependent on each other. Mass supply invites mass consumption. Economics and ecology thus have certain similarities.

Cities as islands of warmth

Cities are warmer than their surroundings due to the 'heat island effect' and the larger the city, the greater the difference. The warming effect is strongest in winter, when heating systems are working at full blast, but it also exists in summer, especially at night, when buildings emit the heat that they have absorbed and stored during the day. During the day the heat of the city can be quite unbearable.

Cities of more than a million inhabitants are typically two to three degrees warmer than the surrounding open countryside, depending on their size and the density of their buildings. If they are situated in a valley or surrounded by hills, they will be even warmer. The warmth benefits the insects, as was evident from the reaction of the moths and butterflies to the particularly hot summer of 2003. This assessment is by no means only true of exceptional conditions, it is simply that these make the effects of the warmth particularly visible. It is generally true for every year and every kind of weather, which is why we see far more butterflies, moths and other insects around the Mediterranean in summer than to the north of the Alps. There would be even more if we travelled to the Equator and searched the tropics, as this is where the greatest variety of insect species occurs. Consequently, large cities actually represent the perfect places in which to investigate the global warming of two or more degrees that is predicted on a vast number of different species. The lives of a great many species in the city could be compared with the same species that live in the country. Concrete findings are always better than the finest mathematical models that must first be tested for their application to nature. Having said this, the less familiar the researchers who use them are with nature, the more popular the models seem to be.

Overfertilized, poisoned land

There is another very important factor that separates town and country, which may not be immediately apparent. It is the extent of fertilization and this, together with the structure of the landscape, represents the main factor that determines the occurrence and abundance of insects, as will be explained in more detail when the specific causes of the decline of moths and butterflies are discussed below. Over the past 50 years or so, the application of fertilizer to land in the countryside has been far greater than it has in cities, which have remained largely spared from its excessive use.

Overfertilization comes about when, over the course of the year, more fertilizing substances end up in the ground than can be absorbed and converted into growth by plants. Problems arise when too many nutrients are extracted, or are extracted too quickly, and the equilibrium is lost, or when the nutrients rapidly move down to soil layers that are too deep for the roots of the plants to reach, or when they are flushed out by the groundwater. To put it in technical terms: over- or undersupply occurs when either more or less fertilizer is applied than can be used as plant nutrients, since it cannot be absorbed through harvest or growth (see Photo 38). The ideal situation would be one that was balanced, but more about this later.

Let us briefly remain with the conditions in the cities. Fertilizing substances tend to reach the cities in three principal ways. First, one fertilizes the garden, to ensure the success and growth of vegetables and other plants and to be able to harvest them. The quantities used in this manner are very slight with respect to the total surface area of the city, especially compared to the mass of manure that is spread on the fields.

We will come to concrete figures shortly. The manure trucks, with their stinking, dark brown slop, do not visit the city parks. However, the wind does carry considerable airborne quantities of it into the cities, above all in the form of ammonium hydroxide that escapes from manure. Most of the fertilizing substances that descend on city land come from traffic and heating systems. With this 'fertilization from the air', we are talking about atmospheric nitrogen that has been combusted and the accompanying particulates that have a fertilizing effect. In the 1980s and 1990s, quantities of 30–60 kilogrammes of nitrogen per hectare landed per year, calculated as pure nitrogen.

Achieving such a quantity of nitrogen fertilizer was a declared objective for German agriculture following the First World War. It was considered a complete nutrient supply for agriculture. Over the course of decades, this quantity was mixed with mineral fertilizer and manure for agricultural nutrient supply and delivered free by air freight before being spread over practically the whole country. With this type of fertilization from the air, nitrogen oxide, NO_x, has a particularly strong fertilizing effect but also causes damage. This is currently the issue with diesel. Its quantity has been reducing in recent years but not rapidly enough. The threshold values for health safety have been, and still are, exceeded too often in the high traffic areas of larger cities. But it would be wrong to think that this NO_x issue was confined to the cities.

A large part of the fertilizing substances that are produced by traffic, combustion and heating are distributed over the whole country and commingled with other atmospheric pollutants that have been released elsewhere. After all, vehicles used in agriculture are almost exclusively powered by diesel. Moreover, the wind blows away what escapes into the air from nutrients and pollutants used in fertilization and soil cultivation in the form of gases and particulates. Further research into where these are blown to is not carried out. Big agriculture is exempt from the principle that the polluter pays, although there is no doubt that the outskirts of villages and forests are strongly affected by this, not to mention nature and water conservation areas in which no direct fertilization by agriculture takes place. In other words, it is by no means just in the city that there is an influx of nutrients and pollutants through the air; the fields and the forests also receive a large quantity, in addition to the slurry and mineral fertilizer that is applied to the fields. However, when it comes to quantity, there is a significant difference between the volume of fertilizer and pollutants in the city and in the countryside.

The volume recorded for municipal areas is essentially far lower than that for the fields. There is a large difference, even if the land in question is not used intensively for agricultural purposes. Only large forests come off better than large cities, but then only if the application of fertilizing substances is properly balanced, since, over time, the forests are also enriched because too little 'export' takes place. Sustainable management of commercial woodland that is geared towards single log extraction and that avoids large-scale clear-felling, where such woodland is made up of pure stands of spruce or pine, actually favours the accumulation of nutrients and the acidification of the forest floor.

Pesticides are now hardly used at all in commercial forest management. Earlier, forests were sprayed with pesticides from aeroplanes in order to protect them from insect pests. In the meantime, the amount of spraying that takes place in the cities has fallen too, and parks are barely sprayed at all. That said, weed killers containing glyphosates are employed in significant quantities on rail and road networks. Then again, the amounts are small when compared to those applied to agricultural areas. It is the land that is being poisoned, not the city. Conventional agriculture is carrying out by far the largest weed and insect annihilation programme that has ever taken place. In comparison, the burning of stubble, field margins and drifts that used to be practised after the harvest was a virtually harmless interference. Living things could cope with that. Since it is visible and conspicuous, stubble burning has been proscribed for decades, but it has been replaced by poisoning, which is invisible and inconspicuous.

Nature-friendly cities

A second key ecological difference between city and countryside is of secondary importance to butterflies and moths, but it says much about the attitude of people. In the city, animals are left undisturbed as far as possible. They are not subjected to a persistent, ruthless pursuit, as in the woods and fields of the countryside. Wild animals are only allowed to live in the countryside in very limited numbers, if at all, if they compete with species sought by hunters and anglers. These groups take it as their natural right to determine in what quantity those animals may exist; they see in this the equilibrium that they have been sent to preserve and see the regulation of nature as their role. The same applies to the agricultural and communal fight against weeds and uncontrolled growth. In commercially managed woodland that is up to 90 per cent planted, and therefore far removed from the conditions of a natural forest, many species are not allowed to grow because they do not deliver any commercial returns. In the state forests, even the edges of the forestry tracks are mown; usually exactly when flowers bloom in early summer and bumble bees are searching for nectar and pollen (see Photo 39). Exceptions are rare and ineffective. Even the state forests, which, after all, belong to the people and are supposed to serve the whole population, are managed according to the principle of maximizing return. The 'credit figures' of the Bavarian State Forests, which are highly publicized as a success, amount to no more than a cup of coffee and a bun per head of the Bavarian population per year. Yet nobody asks whether people actually want such a return, when it is produced at the cost of so much else that the state woods could offer.

I will consider below the extent to which biodiversity is being diminished and destroyed in the state forests when I outline what can be done to stop the disappearance of the butterflies and moths. It is an important point in that context. In the city, on the other hand, and particularly in the larger cities, woods have a quite different, truly public function. There, trees are generally left to grow old and hollow. They do not have to deliver a good yield of timber that can be calculated as a profit to set against costs. Recreational value and beauty take priority over utility and monotony.

The contrast is massive. One sees it as soon as one looks a little closer. In the cities, whatever arrives and can cope with the inner-city living conditions is allowed to grow and live. Control and defence measures are limited to what is strictly necessary and even this is the subject of public discussion. In Berlin, Vienna and other major cities, wild boar and foxes effectively 'take part' in public life. In Scandinavian, North American and East European metropolises, there are other large animals such as elks, bears and wolves. As soon as there is a proposal to fell trees or to build on open spaces in the cities, preservation societies are formed. When Stuttgart's central station was due to be upgraded, the resettlement of the local sand and wall lizard populations cost millions of Euros. Yet nobody cares about the quantities of lizards, slow worms, snakes, frogs and toads that are annihilated every year by land and forest management in the countryside. They are all supposed to be given a high level of protection, just like some of our beautiful butterflies, but this is of no use to these animals, since actions taken by agricultural and forestry management do not count as 'interference with the ecosystem' and agriculture and forestry are released from the strict protective provisions of legislation. Most nature conservationists and lay researchers who study the occurrence and abundance of species come from sections of the public that have nothing to do with the direct use of land. The deliberate killing of animals and destruction of plants without reasonable grounds is frowned upon by the urban population. This attitude increasingly characterizes the difference between town and country.

In summary, this means that cities are (1) much richer in structure; (2) offer better living conditions; (3) warmer than their surroundings; and (4) subjected to far less fertilizer and pesticide than the countryside; also (5) the urban population is much more prepared to take into account the living requirements of animals and plants. Visible expressions of this are the fact that birds in the city are not shy, mammals show themselves by day and do not need to remain hidden in the dark of the night. Moths and butterflies also benefit from the advantages of the city, as the findings clearly demonstrate. The 'decisions' of animals and plants that live in the city ought to be afforded far more respect. They are far more meaningful than the opinions of ideologues who divide things into 'good' and 'bad'. It is not the cities that are bad, but, rather, the countryside that has become inhospitable, and to quite a large extent.

The Inhospitality of the Countryside

In my childhood and early youth, the little village in the Lower Bavarian Inn Valley, in which I grew up and which I have already mentioned a few times, lay in the midst of a diverse landscape. The meadows and fields started on the other side of our garden fence. Three brooks crossed the meadows between the village and the riparian woods by the River Inn. A further brook had its source half a kilometre away from the house, on a terrace. This terrace, an embankment of the River Inn that was formed in the late ice age, separated the slightly higher arable land from the meadows. The whole village was surrounded by orchards. Only a few of them were fenced in and, in winter, roe deer and rabbits came into these gardens. If young trees needed protection from the browsing and chewing, they were covered in a simple wire mesh, like that used for hen and rabbit hutches. In the fields, there were all sorts of grains as well as beet, cabbage and potato crops. Clover fields were interspersed with wheat, barley, rye and oat fields. There was no maize at that point. If I strolled from the house along the meandering field path to the source of the stream, I would go past almost all these different crops. Sometimes, even beans were cultivated. The half-kilometre walk took me past the full range of field crops. When the crops were harvested in late autumn, the network of field boundaries that the rabbits ran along in summer became visible. Partridges and skylarks would nest in this roughly 30-centimetre-wide boundary strip, often surrounded by fragrant, blossoming true and false camomiles with their white crowns and yellow flowerheads, red corn poppies and numerous other wild herbs. In spring, the grey partridge, *Perdix perdix*, would call from the fields with its unmistakable 'chuck chuck chukaa'. In May, the quails would cry 'wet-my-lips' and the skylarks would sing in such great numbers that sometimes I could not tell which of the many birds streaming almost vertically upwards had produced a particular song, before it become mixed with those of the other larks.

Butterflies flew from spring far into autumn. They were especially plentiful in the meadow that backed directly onto our garden because at this time it was full of flowers. There were bellflowers, meadow sage, yellow rattle and many others as well as different species of clover. Some, such as the umbellifers, I found hard to identify, with their round umbrellas of white or yellowish blooms. But I was very familiar with wild carrot, *Daucus carota*, because there was a butterfly that was very interested in it, the swallowtail, with which I was quite fascinated.

From idyll to slurry

In my youthful enthusiasm I even managed to catch one without a net, using just the tent made by my hands. I threw myself onto the ground so skilfully that I was able to cover the butterfly and prevent it from flying away without doing it any harm. My method was a success. With a bad conscience, I pinned the magnificent butterfly and made a small cardboard box for it. I covered it with the self-adhesive plastic that we used to cover our schoolbooks. In this way I could admire it, although it was slightly blurry. I still have it, even though the wings have become detached from the body. My guilty feeling has become less pronounced over the last six decades, but it has never entirely disappeared.

I felt bad because I caught and killed the swallowtail without good reason. In this sense it served as a reminder to me never to kill a butterfly if it was not strictly necessary, even under the guise of science. The fact that I still see a swallowtail from time to time, if rarely, is a poignant reversal of earlier conditions. At the time, these beauties were just beyond the garden fence. Inside it, the cabbage whites would fly and I was supposed to collect and destroy their caterpillars before they could eat too many leaves from our cabbages and cauliflowers.

The meadows behind my childhood home are no longer there: houses were built over them as the village grew. Areas like this that were not particularly profitable were earmarked for construction – areas such as these meagre meadows with their flowers, crickets and butterflies. However, the real clean-up took place across the fields. Twenty years after the idyll of my youth, that is, by the late 1970s and early 1980s, the area was unrecognizable. Its diversity had been transformed into homogeneity. The great transition, in which the meadows were converted into arable land, took place in the 1970s. Ten years later, the village had become an island surrounded by a green ocean of maize. The same could be seen along the whole length of the Lower Bavarian Inn Valley and beyond, since the voracious 'consumption' of new land by large-scale agriculture in this part of southeast Germany was truly worthy of its name. Even today, the districts of Passau, southern section, Rottal-Inn and their surroundings make up one of the German, if not European, centres of maize production. The maize seems to creep up between the hillsides; it has already spread throughout the pre-Alpine hill country, but also encroaches on the fringe of the Alps and up the Alpine valleys. In 1960, maize production covered only a couple of thousand hectares in Germany; now (in 2018), it covers 2.5 million hectares, that is, 1,000 times more land. Thanks to the energy revolution (i.e., the use of maize as a biofuel), the area of maize cultivation had already risen to 1.5 million hectares by the late 1990s. Maize cultivation was heavily subsidized by the state. With the arrival of biogas from biomass, farmers rapidly changed their crop portfolio. They became energy farmers, while still maintaining all the privileges and public subsidies of land farmers.

At the same time, another fundamental change was taking place in the agricultural sector. Livestock farming was, to a great extent, moved from pastures to stables where water and manure were removed to produce slurry. Given the extraordinary size of the cattle population in Germany, the volume of slurry produced is hard to imagine: 310 billion litres per year. A single administrative district with a high cattle population but only around 100,000 people produces as much or more sewage (= liquid manure or slurry) from livestock farming as the 3.5

million people of Berlin. For the whole of Germany, with its 83 million inhabitants, the quantity of liquid manure is two to three times as much, depending on which excretions from livestock farming are used for the calculation. Let me add a couple of concrete figures from the German Federal Statistical Office for the year 2017. The agricultural animal population in Germany in that year was made up as follows: 12.3 million cattle, of which 4.2 million were dairy cattle; 27.6 million pigs, of which 1.9 million were breeding sows; 1.6 million sheep; almost 778 million chickens (known as pullets or broilers) and a further 40.6 million laying hens. The 1.3 million horses are only a small minority in comparison with these huge figures, and only a few of them are still in use in agriculture. The vast majority are riding horses and are thus classified as 'pets', like dogs and cats. With regard to the pigs, it should be noted that their number does not actually represent the total annual population, since most pigs do not live to be more than six months old. To arrive at a total figure for food and the resulting manure, a significantly higher number would therefore have to be used – for the broilers, one that was several times higher. From this we may reach the conclusion noted above, that animal husbandry produces several times as much sewage as all the people in Germany. Almost all human sewage is treated and consolidated by means of very effective and expensive sewage treatment facilities. In contrast, the hundreds of billions of litres of slurry end up directly on the fields, with huge consequences for nature, not only for plants and animals but also for the quality of air and ground water. After all, these vast quantities of animals must be looked after, if they are to produce corresponding profits. Sometimes this includes the use of medicines and other additives that inevitably place a burden on nature and our environment. All of this has been reported frequently and in vain; the agribusiness seems to be immune to its significance.

Monocultures and changes to the ground-level microclimate

This immunity has been transformed by politics to virtual unassailability. Many people have already complained about this too, without any success. The endless chain of scandals in agriculture is still not enough to bring about fundamental change in the concerns of the overall population and their future. As it was in feudal times, today, too, he who owns the land has the last say.

But enough of this lamenting, which is futile anyway. The point of the city/country comparison was to determine the main causes of the disappearance of the moths, butterflies and other insects as well as the plants, birds and everything wild that lives (or lived) in the fields. What can we take away from these descriptions? Colourful blossoming meadows that we remember from our childhood and youth are hardly a persuasive objective comparison between past and present. Like all memories, they are filtered and coloured by the years and decades of our lives. It would be better to examine more closely the uncontested and verifiable criteria governing the differences between town and countryside with regard to moths and butterflies.

We need to try to understand present circumstances and draw whatever conclusions present themselves without the aid of any recollections, however they are formed, and in fact this is quite simple. The issue of overfertilization is best suited to this, since there are ample areas with different quantities of plant nutrients in the ground. The quantities that are directly applied to the ground or the air can be measured quite accurately; in any event, more accurately than the plants that react to them. The available nutrients cannot be immediately absorbed and converted into plant tissue and thus the growth of plants takes time. How the plants actually react has been scientifically researched in depth. The fundamental process is photosynthesis, by which nutrient salts such as nitrogen and phosphorus compounds are combined with water and carbon dioxide from the atmosphere to produce plant material. Through this process, a quantity of oxygen is released that corresponds to the amount of carbon dioxide that has been absorbed. Assuming an appropriate supply of water, the performance of the plant depends on the mineral substances necessary for growth. We refer to these as plant nutrients or fertilizer. Ultimately the growth and success of all plants depends on fertilization.

This is true for individual plants as well as for the larger plant community. For moths, butterflies and other insects, it is vital that vegetation, that is, the community of all the plants in a given area, grows thicker and denser the more fertilizer is applied. Fertilization promotes growth and this is the objective of agricultural plant production. This also applies to forests, managed woodland and the growth of trees. In managed woodlands, we can observe what is generally hidden from human view in the meadows: the competition of trees with each other. If a young group of trees grows, regardless of whether it is planted or grows naturally, and is allowed to develop without the interference of humans (which is rare enough), more and more of the saplings and younger trees will die out. After several decades, only very few will have survived, and this is clearly visible in older stands of trees.

For example, if there is a young growth of willows whose seeds have been washed to a new island in the middle of a river, it will start off with thousands of seedlings per square metre. About 50–70 years later, only one of them will be left, and not on a square metre, but on 20–30 square metres. Tens of thousands of its competitors will have been sacrificed for its survival, and this is the natural condition. Yet competition from the same species, fierce as it is, only accounts for one part of the losses due to displacement. Many other plants fall victim to this competition and fail to survive. The young plantation was too vigorous; its supply of nutrients too good.

If, on the other hand, the nutrients are scarce, growth will be slower, and vegetation will remain sparse for a longer time. More sensitive, less competitive species will survive, if not permanently, then at least for years or decades. The vegetation will thus look more diverse, while the pure stand of young willows resembles a monoculture. In a monoculture, only those insects that are specially adapted to this species of plant, or those with caterpillars or larvae for which this species is an appropriate food plant, are likely to occur. Many species, on the other hand, support many users. These are simple relationships that are readily understandable even without special scientific knowledge; the direct consequences are also readily apparent. If a single species or a few species of plants have very good growing conditions in the relevant area, they will thrive and produce a dense thicket. This leaves no gaps for other, weaker competitors. Heavy fertilization thus promotes species-poor plant communities that multiply quickly. Diversity is associated with scarcity. As soon as living conditions improve, certain species will win the upper hand at the cost of many others. The higher the level of fertilization, the lower the level of biodiversity: that is the inevitable consequence of this relationship.

But for vegetation, there is a further factor that has a particularly powerful effect on the creatures that live off it. Vegetation becomes moister and cooler the more luxuriantly it grows. The microclimate will be much cooler and wetter, particularly near the ground, than in loosely growing plant communities where the sun can reach the soil surface and warm it up. As the ground-level microclimate becomes cooler and damper, many insects will no longer be able to live there, even if this is where their food plants grow. This is because, in their feeding stages, as larvae and caterpillars, they need warmth and sun for development, just as plants need light.

The cooling of fields and forests

For a great many insects that live in meadows and fields, their living conditions will deteriorate in this manner long before their food plants completely disappear. The vegetation not only grows too dense but also grows too quickly in spring. For the occurrence and abundance of moths and butterflies, especially those species whose caterpillars live in the open, on the ground and in vegetation close to the ground, the microclimate is usually far more decisive than the 'official weather' as determined by the weather stations. A fine warm day in May with an air temperature of 25°C can be much too cold and wet for newly emerged butterfly caterpillars in a meadow where the grass has already grown knee-high. I would like to stress that what is critical, not only for butterflies and other insects but also for the small birds of the field, is not what the weather stations measure and what is meteorologically considered to be 'weather and climate', but rather the ground-level microclimate. Due to the overapplication of fertilizer that has taken place in our fields over the decades, this is considerably cooler and wetter than in the nineteenth century, although that period was cooler and wetter in meteorological terms. This cooling effect starts to take place long before the actual disappearance of the food plants of the caterpillars – this is also worth re-emphasizing.

Our woods have also become cooler and damper because they have been so massively fertilized from the air. They are denser than ever. This is particularly noticeable in the riparian woods, which have really proliferated beyond the dams despite no longer receiving new nutrients from flooding. The relentless application of fertilizer from the air has for decades provided them with more nutrients than the irregular flooding from the unregulated rivers once did. The problem of overfertilization is also apparent in every protected area that is left to itself. The vegetation on these unmanaged areas grows ever more luxuriant and with unexpected speed. While the effects of earlier land plot-divisions still persisted in the 1960s and 1970s, the riparian woods along the River Inn were still richly structured, but this all changed with the end of coppicing. The riparian woods grew much denser than in earlier times, when dried-out reed canary grass and other tall grasses were collected from the floodplains for use as straw for cattle stalls. This winter use of straw, which prevented the development of a thick, matted growth of vegetation along the ground, provided favourable conditions for snowdrops, squill and oxlips. It promoted biodiversity of plants in the undergrowth in general. After this very labour-intensive form of forest management was abandoned, since straw was no longer required for slurry-producing water and manure extraction in the cattle stalls, the spring flowers disappeared from these areas, either in large part or completely. Some species of shrub and tree also became less common, displaced by the more rapidly growing white willow and planted hybrid poplars. The same is true for both the riparian woods and the grassland: the stronger the growth of the plants, the stronger the cooling effect brought about by the evaporation of water – that is, transpiration.

Increased growth reduces the abundance of moths and butterflies in the riparian woods

The decline in quantity of moths and butterflies in riparian woods, especially pronounced in those that have been left unmanaged for decades since the end of coppicing, is based on the cooling effect brought about by the increasing concentration of vegetation. This is revealed particularly in the fact that the large Lepidoptera such as hawk-moths and certain other species have decreased far more than average. Large caterpillars require more warmth than small ones do for their development. In the commercially managed woodlands, similar developments are in progress, especially when clear-felling is completely avoided because it is considered 'ecologically unsound'. What exactly 'ecological' means and how it is evaluated is highly debatable. Today's practices are based on the ideology of a time when clear-felling was condemned. Certain very influential personalities from the forestry community shaped nature conservation. Their one-sided perspective, which focused on the (commercially usable) stock of trees, did not consider the many animal and plant species that thrived in the early stages of forest development. Such spaces would once have been created by forest fires, but these are no longer allowed to take place; on the contrary, they are extinguished as quickly as possible with the most modern technology and at immense cost, for example using fire helicopters. As a result, these early stages of forest development no longer exist. They are even rarer than areas where the trees are left to get old and unused stands of trees collapse. The commercially managed woodlands of today are artificially kept in a state that is particularly productive for timber, the only permitted changes being the gradual transformation from coniferous to deciduous woodland, 'in light of climate change'.

Yet no new forest grows on the skid trail; not much sunlight reaches such tracks. Below, the ground remains wet and cold. The heavy timber-harvesting machinery presses deep furrows in the forest floor. Although they are clearings of a sort, these aisles created for the extraction of timber are not suitable regeneration spaces for all the 'useless' plants and animals of the forest. The decrease in abundance of many forest species among the moths, butterflies and other insect species is thus essentially based on the same factors that influence life in the open country: the acceleration and increased density of growth, through overfertilization and the homogenization of vegetation.

Boundary ridges in the fields and meadows: a supportive network

At this point, I think a reminder of the differences we noted earlier between town and country would be appropriate. Prior to large-scale land consolidation after the mid-twentieth century, small structures in the countryside guaranteed a high level of biodiversity in the fields. Field margins and hedgerows were more open, sunny and less densely grown than the fields, since the farmers took care to spread manure and slurry only on the field itself, in the places where the crops needed the fertilizing nutrients. More nutrients were taken from the soil through the harvesting process than were put back through the traditional application of fertilizer. Field margins and verges, hedges and drifts were thus not as 'overgrown' as those that survived the land consolidation; they remain here and there as a symbol of what has been lost but are now overfertilized along with the fields. Formerly, the tightly meshed network of the field margins remained sunny and dry.

Furthermore, these structures created boundaries, which in turn promoted species diversity. In the ecological science, the significance of these boundary or 'edge' effects are well known. Habitat boundaries naturally create gradients. Such gradients no longer exist where there is just a deep furrow dividing two fields or where neighbouring arable areas are simply contiguous, since the one that produces maize and the other that produces wheat belong to the same business anyway. These boundary effects only survive where the structure of the landscape resists such standardization, such as in the fields of the Swabian and Franconian Alb and in a few other areas of central Europe and Britain that have a morphologically very varied terrain. In fact, it is these areas, together with the mountainous regions, in the strictest sense of the word, that have, up to now, lost the smallest number of species – relatively speaking (rather than in absolute terms), as the records for the nature reserve near Regensburg showed. Dr Andreas Segerer and his co-authors established this with their comparison between earlier species counts and recent records: the number of species of butterflies had sunk from 117 in 1840 to 71 in 2015; that is, by almost 40 per cent, even though the slopes had not been managed.

There is a corresponding example to illustrate the sharp decrease in species numbers from the findings from northwest Germany that were brought together in the Krefeld Study.* There, starting in the early 1990s, insect populations were measured using a different method (see pp. 202 et seq.) and a decrease of more than 70 per cent of biomass

---

* A study, published in the journal *PLOS One*, carried out by dozens of amateur entomologists in Germany, using standardized ways to collect insects, between 1989 and 2016. More than 1,500 samples of flying insects were captured and recorded at 63 different nature reserves.

was established – even in protected areas like nature reserves. This broad similarity is an illustration of the ecological principle that, where populations decrease in this manner, first overall abundance decreases, then more and more species disappear completely. In the investigations carried out using light traps from the village outskirts close to the fields, almost identical results were obtained, with an average loss of half the number of species (see Diagram 8 above) and a loss in quantity of over 80 per cent. There is no doubt that the decline in moths and butterflies took place on a large scale. Sites investigated included conservation areas and remote areas that were not farmed very intensely. Large-scale studies in richly structured landscapes did show smaller, yet still surprisingly large, changes in the ranges of species. For example, Walter Sage established that in southeast Bavaria, 73 species of moths and butterflies had totally vanished from the area since the mid-1990s, 19 species had become rare, 92 species had become significantly rarer, and 117 species were hardly ever found and would need to be considered threatened; furthermore, a slight increase could be confirmed for only 8 species and 13 new species had entered the region as a result of the expansion of their range.

For an area with plenty of woodland, riparian woods, large bodies of water, nature reserves and a low density of settlement, with a favourable climate, these numbers are as shocking as the insect decline reported by the Krefeld Study and the decline in butterflies near Regensburg. However, the areas with by far the highest losses in moths and butterflies were the large agricultural areas, the fertile plains – or whatever the regional terminology is. These are the areas – not the big cities – where nature is being obliterated. The same is true for the forests and woodlands. If these areas contain many habitat structures in the form of hillsides, gorges, ridges, or water bodies, and their tree stands contain a diversity of species, then the species range of the moths and butterflies will have been preserved. However, if they are homogenous when it comes to species and the age of the trees is very similar, as in many commercially managed forests situated on flat land, then species decline will have affected them too. The parallels with the results from (large) cities are evident.

'Infilling' and 'compensating areas'

The larger the cities, the richer they are in structural diversity. Over the course of the last 50 years, they have profited from town planning that is increasingly oriented towards human quality of life. The management of city parks, inner-city copses and water bodies is now geared towards their recreational function. The practice recently has been to refrain from excessive cultivation and maintenance, in part because costs can thereby be reduced. Considerable areas of green open space are no longer kept in the perfect condition of traditional English lawns and mowed several times a year. Instead, flower meadows are allowed to develop and butterflies may fly over them. However, over the last 10–15 years, a practice that seems to work in the opposite direction has been introduced in Germany, namely 'infilling' the open spaces with new houses and buildings. Available, unused but publicly accessible open areas are being built on, so that the 'evil city' does not gobble up any more of the 'good country' with its rampant growth.

Internationally, the thinking has long been quite different. Elsewhere, it is thought that every city dweller should have access to public green spaces that can be reached on foot. In London and other big cities, appropriate specifications have been made and new open spaces created. Clearly it is only Germany and a few other countries that still cling to the old idea that has been comprehensively rebutted elsewhere. Here, the quality of life of city dwellers is sacrificed to give priority to maize fields. Evidently this way of thinking belongs to the Green ideology, which suggests that the city must be bad and should be made so again, if it has turned out to be too 'good'. A highly significant side-effect of the prescribed infilling is the exorbitant rise in land prices. In cities with a high level of residential influx, prices are no longer affordable unless investments are made by large corporations. Any open areas still available are converted into fountains of gold for the city treasurers. For financial reasons, it is virtually impossible for any town to resist building on undeveloped land. This is another effect of the politics of infilling that should give food for thought.

Urban infilling follows the same pattern created by the reorganization of the countryside into large, standardized, arable areas. Small structures were destroyed by field consolidations, and field areas were enlarged as several of them were merged together, homogenized and used far more intensively. Ultimately, the so-called 'ecological compensation areas' that exist in Germany are subjected to less intensive management, but this is still on much too small a scale.

The idea behind the ecological compensation areas is good. Large changes ('interventions') that impair species diversity and natural processes should be compensated for with areas that are reserved for nature and not managed. Together with nature reserves and organic farming, they are supposed to provide a counterweight to the losses caused by the industrialized conventional agriculture. However, it is highly questionable whether their 'wins' for biodiversity truly compensate for the losses caused by urban infilling. The fact is that extraordinarily species-rich biotopes have been, and continue to be, sacrificed as a result of inner-city building development. Nowhere else is there such a richness of species in small areas as on fallow or abandoned terrain that is distant from land that is subjected to intensive agricultural use.

Quantitative considerations should be included. The total settled area in Germany occupies about 12 per cent of the land surface. Agricultural areas comprise 52 per cent. Of this, only a few per cent, perhaps two or three, are organically used in such a way that truly promotes species diversity. The areas set aside for ecological compensation are only a tiny fraction of the whole. So, for example, if redensification in a city leads to a loss of 2 per cent of its area, then, to compensate for this, more than 2 per cent of agricultural area outside in the countryside should be set aside for nature. Realistic? Certainly not, since infilling does not carry any obligation to compensate – any more than the intensification of agricultural ground use does. Compensation only needs to be provided in cases where building work takes place around transport locations, such as road constructions, railway stations, airports and the like. In these areas, compensation measures must be carried out even if the building works themselves improve the condition of the natural environment and are not intended for commercial use. Accordingly, the extension of a former, in places completely silted-up, branch of the River Inn to ensure permanent flow had to be compensated by the planting of new trees, since trees and bushes had grown up on stretches that were now only flooded at times of extremely high water. The fact that a whole range of aquatic animal and plant species would now live in the renewed waterways, and that fish would have new bypasses to swim upstream that were previously blocked off by dams, counted for nothing. The fact that these absurd regulations do more to hinder than to promote is obvious, yet they are not amended. This can give the impression that their main objective is to make money from the bodies responsible for construction, or to force them to carry out tasks in order to delay or circumvent projects that have been rejected by certain sections of the public. This is how 'species protection' is carried out.

In any event, it would be far better for people if the open areas within cities were not built over, but any necessary new construction were to take place instead on the maize fields at the periphery of the city. This would retain quality of life for those in the city as well as promoting natural biodiversity. Cities should follow the pattern of organic growth and be able to develop peripheral centres while avoiding excessive density in the core areas. Fractal structures are not only superior in mathematical or physical models, but also in real life: colonies of animals and density-dependent developments in the plant kingdom follow the same pattern. Diversity and species diversity thus come about quite by themselves.

To reiterate: many more animals and plants can live and grow on a well-planned supermarket carpark at the edge of a residential area than on a similar area that is subject to intensive agricultural use. The extensive losses of wild animals and wild-growing plants that we are discussing did not take place as a result of construction and settlement activities. Construction companies are nowhere near the worst culprits when it comes to bringing species close to extinction. Whether nature conservationists like it or not, construction and settlement activities cannot be made responsible for a quantifiable share of the losses to biodiversity, nor for the fact that birds and mammals, butterflies and other insects are disappearing. Measures taken in the name of nature and environmental conservation probably cause more losses than construction and settlement activities. The survival of the species that remain will depend on the type and intensity of the means by which they are managed. Structural factors will form the restrictive underlying conditions, although this is also all too frequently overlooked.

The 'nutritional condition' of the landscape

The effect of land management has become intensified through the application of fertilizer and pesticides. As was noted earlier, the city comes off far better than the countryside in this respect. The overfertilization to which the land is subjected has a nation-wide average in Germany of more than 100 kilogrammes of (pure) nitrogen per hectare per year. However, two to two-and-a-half times this amount accumulates in those areas where arable farming and animal husbandry are practised most intensively. Since the records for quantities dispersed by air to town and countryside are approximately the same, what is relevant for the comparison is essentially how much manure and mineral fertilizer are spread on the fields. The quantities are so high that there is practically nowhere in the German countryside where nutrient-poor conditions exist. If this was addressed artificially through soil loss, the nutrients would still accumulate over time and produce a phenomenon that was a rarity 50 years ago but that has since become the norm: eutrophication (from the ancient Greek *eutrophos* = well-nourished). This technical term refers to a degree of nutritional availability that exceeds the actual needs of the vegetation. The opposite, oligotrophy (from the ancient Greek *oligo* = little, slight, small) is a condition of the soil that has become decidedly rare in the countryside. This is because even if the soil is not fertilized directly, it receives such a plentiful supply of nutrients by air that any nutritional deficiency in the soil is short-lived. This need not be the case.

Until the 1970s, our water bodies had become eutrophic through the influx of untreated wastewater. In simple language, they were heavily polluted, in places catastrophically, and this had many negative consequences, especially with regard to drinking water. Many lakes and rivers became unsuitable for swimming, and fish died. The inordinate discharge of sewage had to be stopped. This was only possible by means of thorough purification in highly efficient sewage treatment plants. Billions were invested in the treatment of wastewater. This is rightly considered to be a special and successful instance of environmental protection, but it had some unusual repercussions that have been making headlines for years: many lakes had become too clean! The earnings from fisheries were thus heavily reduced. The demand to put more fertilizer back into the lakes is understandable from the perspective of commercial fishing, and yet it is unrealistic. The water quality that had been improved at such high financial cost could surely not be impaired simply for the convenience of a few users. Water quality is connected to nutrient content. The best water quality is achieved when there is nothing living in it anymore, because then the water is quite safe to drink. This is the logical, stated goal of wastewater treatment, however unattainable it may be. Unattainable, because purification through sewage treatment plants only addresses human sewage and not the slurry produced by agriculture. As noted above, its volume is several times greater than household wastewater. The resulting pollution affects mainly groundwater and streams in the fields.

The most favourable and ultimately prudent situation would be somewhere in the middle, with an equal relationship between availability and use of nutrients. This is known as the mesotrophic state (from the ancient Greek *meso* = medium, centred). If this could be maintained over a longer time period, then the water bodies and soil would remain productive, and the ground water would not become polluted. However, it is a transitional or transient state. States that are stable in the long term are nutrition-rich eutrophy and nutrition-poor oligotrophy, and this is also true for soil.

For hundreds of years, agriculture withdrew more nutrition from soil than it returned in the form of farmyard manure and swill. The soil became impoverished and yields decreased. The 'good old days' of the extraordinary biodiversity of the nineteenth century were mainly times of famine. Millions of Europeans felt forced to emigrate. The deficit could not be remedied without artificial fertilizer, since the agricultural sector could not nourish the population that had increased so sharply in the wake of the industrial revolution. This deficiency in the soil was to last until after the Second World War. It was only in the 1970s that the nutritional shortage in the soil was balanced out in Germany by the mass application of artificial fertilizer. But this only lasted for a couple of years and the balance was only achieved by taking a rough average of differing conditions across the country. Just as in the lakes and rivers, the balance of the mesotrophic state could not be maintained. It took less than a decade for this level, favourable to the whole environment, to be far exceeded. The eutrophic condition had been reached and had become stable. Since then, the application of nutrients to the soil has taken place at (much) too high a level. Despite high rates of withdrawal through profitable harvests, much remains behind in the soil, with corresponding losses in the groundwater, the quality of which decreases steadily with increasing fertilization.

Drinking water can no longer be taken directly from groundwater in many places, and in some places it has been unsuitable for this purpose for a long time. Given the slow speed at which bodies of ground water are renewed, its current polluted condition will last several decades more. In order to 'rebalance' certain soils, especially heavy, argillaceous ones, no more fertilizer should be applied for years or even decades. This is one of the conditions that makes the conversion of agricultural businesses to organic production so hard. The environmental and nature conservation sectors were too late in recognizing the consequences of the overapplication of fertilizer for the plant and animal kingdom, although these had been researched and scientifically demonstrated long before. The agricultural sector refuses to acknowledge this and introduce countermeasures. New water protection areas face the same kind of resistance as the construction of new motorways. The eutrophication of the fields actually developed in a similar way to that of the rivers and lakes, but in a much less public manner, since the changes that it caused were only revealed very gradually. They are still invisible, just like the poisoning of the environment with the insecticide DDT in the 1960s.

The disappearance of the cockchafers

DDT was applied very widely in Europe – certainly not in the quantities used in tropical and subtropical areas, but quite heavily enough to combat pest insects. Population explosions of butterflies and moths with caterpillars that caused damage in forests also caused problems. As an insecticide, DDT was employed in a variety of ways, but it was also partly responsible for the collapse of the cockchafer population. Until the late 1970s, there used to be a three-year cockchafer cycle in southeast Bavaria, as elsewhere in the warmer regions of central Europe, while in the cooler regions, a further year was required until the cockchafers were ready for their mass flight at the start or in the early weeks of May. Swarming would begin in the evening hours in suitably warm spring weather. Usually, it would only last two or three evenings. The beetles would already have emerged from their pupae in the preceding autumn and would have spent the winter and early spring in a burrow until the increasing warmth of the soil towards the end of April gave them their signal to emerge. Such temperature regulation enabled them all to swarm simultaneously.

These large brown beetles would then climb up, buzzing, all over our garden and the adjoining meadows at sunset and fly around before stopping at a tree or a shrub that was silhouetted against the horizon. Preferred targets were chestnuts, where the leaves were just emerging, but they also headed for alders and other trees with fresh green leaves and fed there. In some years, their consumption was so great that the trees almost lost their entire first generation of leaves. If the weather in April was persistently cool, the flight of the beetles was delayed and the young foliage of the trees would be slightly further developed before being eaten. Cockchafers were considered pests. However, they were controlled in a manner that, in hindsight, appears wonderfully naive. Schoolchildren were sent out to the trees at the start of lessons to shake the cockchafers out of the small trees and collect them. The beetles were then fed to the chickens that existed in large numbers in the village at the time and were naturally free range. Their eggs would take on a curiously nutty flavour. And since the swarming would only take place over a limited number of evenings, the general opinion was that the cockchafer control had been successful. After this, there would be no cockchafers for the next two years, or at best just a few individuals, which flew around aimlessly.

In the third spring they would reappear. Nothing would have changed in their quantity, and the practice of three years previously would be repeated with the same results. The cockchafer larvae, the cockchafer grubs, needed this long until their development was complete – they could not manage it in one or two years. The reason for this three-year cycle was evidently unknown to the rural population, although the cockchafer beetle itself was the most well known of the beetles: the quintessential beetle. Its larvae, prior to pupation fat, cream-coloured grubs with brown head capsules, destroyed some plants in the garden by eating their roots. Their effects could also be seen in grasses, if larger areas became brown, again because their roots had been eaten. Yet the fact that the cockchafer flights ended so suddenly that there were only a few individuals after 1978, the last year that saw a significant population flying, and none at all from the mid-1980s, was nevertheless perplexing. Those cockchafers were not supposed to become that rare or even die out completely. In Germany, a popular song, 'There are no cockchafers anymore', was released in 1974 by Reinhard Mey. Already by then, the disappearance of the cockchafers was imminent, and they would not return – not even the common cockchafer, *Melolontha melolontha* to give it its proper scientific name, although its close relative, the woodland cockchafer, *Melolontha hippocastani*, still flies in its three- or four-year cycle in some places. Its habitat is not made up of open fields and gardens, like the common cockchafer, but bright deciduous forests. The two should therefore be distinguished. Reinhard May was singing in protest against the multistorey carparks and against the growth of cities and all the changes that they brought, unaware that his song would take on meaning in another, far larger context. But why did the common cockchafer disappear? Why at that particular time, and why did the factors that caused its disappearance not affect the woodland cockchafer, which is otherwise so similar that they are hard to tell apart? And how is this relevant to the disappearance of the moths and butterflies?

The cockchafer grubs obtain their nutrition from roots; accordingly, large populations can cause significant damage. As roots are not particularly nutritious, the development of the grubs takes several years. Roots are also hard to digest, and for this reason the grubs need assistance from digestive bacteria that can make more out of this foodstuff, which is very rich in sugar and cellulose but otherwise poor. The protein content is not sufficient to enable the female cockchafer beetles, once emerged, to develop eggs in the quantities needed for successful propagation. The remainder of the protein that they require is therefore absorbed in the so-called 'maturation feeding' stage when they feed on young leaves. It is also a prerequisite for the mating of cockchafers, hence the importance of the swarm at the exact right time, when fresh and nutritious greenery is available. These requirements for cockchafer breeding underscore the constraints brought about by successful adaptation. Success can actually be limiting in this respect. Similarities with the ermine moths are evident.

But there is a still further effect, which may not be apparent here: that of the symbiosis with microbes that contribute to the utilization of food. If it is compromised, the whole complex relationship falls apart. Now there are strong indications that suggest exactly this. At the time, chemicals used to protect roots from fungal infestation were being introduced by agriculture. Fungicides are particularly significant in crop protection because fungal diseases often cause greater crop failures and profit losses than all other pests. At around the same time, that is, in the 1970s, the widespread conversion of meadows to fields took place, together with an increased use of fertilizer, especially slurry, with which fields are still temporarily inundated. However, this treatment did not extend to the forests, least of all broad-leaved forests, which were not affected by either the root-protecting chemicals or the floods of manure; nor were they largely reduced in size. Sparse forests with poor soils, moreover, do not grow as thickly and luxuriantly as densely packed commercially managed woodland in areas with high precipitation and correspondingly higher levels of air-borne fertilization.

The differences in reaction between the two species of cockchafer can thus be clearly understood. The collapse of the common cockchafer populations corresponds to the introduction of fungicides and the transition to overfertilization. This also makes the ground cooler. The swarming can therefore 'lose its rhythm', since the 'warmth signal' is no longer simultaneously affective across a wide area of land. Speculations? Yes, certainly, but they are well grounded. They are based on similar developments with butterflies and moths, in particular the dark-arches moth, *Apamea monoglypha*. The dark-arches moths, known as 'root owlets' and 'root-eaters' in German, whose caterpillars are supposed to have caused damage to grain roots, as noted in 'Koch', disappeared from light traps at the village boundaries at the same time as the common cockchafers. In the city, and in larger towns in Germany, in contrast, the 'root-eater' continued to occur at normal frequencies, while the common cockchafer has also survived, up to now, in residential areas. Here they do not cause damage, and indeed, a trend indicating recovery is discernible from the records of (the very warm) May 2018. But this may be premature: the coming decades will show whether the cockchafers are returning or whether apparent population spikes like that of 2018 can simply be ascribed to the weather and have no lasting consequences.

The turning point for our farmers: the 1970s

Let us return to the last days of the cockchafer swarms. For about a decade, the fields still had buffers, that is, deficits that had built up in the ground over the centuries. But the motto for fertilization was 'more is better' and this was practised not only on the fields, which had been prepared for high demand and mass production and unified through land consolidation, but also privately, in gardens. Nitrophoska, the nitrogen, phosphorous and potassium fertilizer that contains the primary nutrients required for plant growth in appropriate proportions, was the miracle product that made everything sprout and grow.

Crop yields grew. Production rapidly overtook demand. Grain mountains and, in livestock farming, butter mountains, milk lakes and pork mountains characterized the transformation to surplus production, which had to be managed in the Common Market at great cost to avoid prices falling. A vicious circle had been set in motion: as a result of the concentration by agricultural businesses on a few, and increasingly just one, field crop and the enlargement of production units, costs were lowered, but at the same time competition between farmers was enhanced. The surpluses could not be reconciled with demand. The result was a massive decline in farmsteads. More and more farmers had to give up because the areas that they managed were too small to withstand the pressure from competition and to carry the enormous costs of the machinery and the monocultures required. In just 30 years prior to the turn of the century, approximately 90 per cent of farmers in Germany gave up farming. The remaining 10 per cent survived as businesses because they received area-related subsidies from the state or from the EU agricultural budget. This is a state-controlled command economy. In practice, the public has already bought off the farmers' land several times over with the subsidies.

And what about the pesticides? This question is surely asked with increasing impatience. After all, it cannot possibly be the case that the huge losses in moths, butterflies and other insects are due solely to overfertilization. This critical question is justified, but at the same time difficult to answer, since much too little research or investigation has been carried out in this area: most of it must be inferred by making comparisons. As has already been extensively described, the decline of moths and butterflies is particularly evident at the village outskirts, but less so in habitats further from the fields. This finding can only mean that the effect of the pesticide sprays extends beyond their target areas. The numbers of moths recorded in the village outskirts decreased especially sharply in the month of June, by 89 per cent as compared with the 1970s, when this area was surrounded by meadows that had never been sprayed. In July and August, the numbers did not sink as far. In total, there was a decline of 75 per cent. In larger gardens that were far enough away from village edges and in the cities, there were, by contrast, no losses in number, and species diversity also remained unchanged.

In towns and gardens, one can still find those species of moths and butterflies that used to live in the fields. If this were not the case, then the list of Lepidoptera species threatened with extinction would be about 100 items longer. Naturally, they do not occur in the residential areas in such great numbers as in those days when the main flight came from the fields, since the suitable urban areas are simply too small. It is therefore highly likely that the effect of the pesticides went beyond the land parcels that were treated with them. It is important to review not just how much of the crop protection products employed by agriculture end up in the ground water; a thorough review should also find out what is carried to the adjoining, non-sprayed areas and to the streams, and what quantities are carried into the villages. The characteristic stench often indicates that it is present.

The Krefeld Study

The Krefeld Study (see p. 190) demonstrated just how badly protected sites and nature reserves had been affected. In the second half of 2017, its results were discussed in public at an unprecedented level. The findings even found their way into the coalition agreements of the newly formed German government. The awareness that things cannot be allowed to continue like this is rising: the study showed that insect biomass had decreased by over 70 per cent since the 1990s, even though the places where these numbers were recorded were all either in nature reserves or in other areas not used for agriculture. This decline can only have been caused by the combined effect of fertilizers and pesticides. The application of fertilizer, which in this case came above all from exposure through the air, strengthened plant growth and caused the principal effect of colder and damper living conditions in the zone close to the ground. This expelled those species that need warmth and sunlight. Others that could cope with the increased density of plants and should thus have become more abundant did not compensate for the losses. The Krefeld Study was concerned with the mass of insects, more precisely their biomass, or live weight. This, not the mere number of insects, had decreased by almost three-quarters.

Better fertilized biotopes should have produced more insects, that is, the numbers that were caught should have been higher, since the protected areas in which the investigations were carried out were not used for agriculture. Yet, despite strong fluctuations from year to year and differences between the individual traps, the quantities had decreased by about a quarter around the beginning of the 1990s. Since the malaise traps used were placed in quite open areas, the insects that were caught are, to a great extent, probably associated with this type of habitat. Species from dense forests cannot be so easily caught with this type of trap. In practice, they are constructed like fish traps: insects flying into the trap end up in an open gateway, from which they travel along a passage made from fine mesh that gets narrower until it culminates in the actual trap, which consists of a container filled with preservation fluid at the furthest, highest point of the apparatus. The insects are removed from time to time, classified into species or groups, accurately weighed – in the case of the Krefeld Study – and their total weight, their biomass, determined.

The contributory effect of pesticides from crop protection products that have been applied in the area thus becomes highly likely. It would be instructive to discover whether and how the composition of the range of species of insects recorded from the malaise traps has changed in the last quarter century.

The findings for the surroundings of Krefeld are largely consistent with the results of my own investigations in southeast Bavaria, a good 500 kilometres away. In terms of timing, they cover a much shorter time span, from the second half of the 1990s to 2013, the period during which my records focused on Munich. In ecological terms, the Krefeld findings supplement my own by considering nature reserves surrounded by open land that is used for agriculture. With respect to their findings, they therefore lie somewhere in the middle, between the extremely high decline of insects in the intensively used agrarian landscape and the moderate level of loss in the forests. If one adds the residential areas, one arrives at a relatively complete series with a collapse towards the fields: largely stable populations in the city, smaller populations in smaller towns and villages, high losses outside these, even in protected areas, and only minimal residual populations of insects in the conventionally used agrarian landscape. As bizarre as it may seem, it is not the metropolis that spells the end of nature, but the maize field.

A subsidy system without an exit mechanism

But why should that be so surprising? Only maize, rape-seed oil, wheat or some other crop grows on the areas of land used for agriculture; otherwise nothing. The plants that have established themselves in this space from the time farming began are weeds that have been controlled with hoes, machines and, finally, poison. Animals that live from these crops are pests. They are also controlled using poison and attempts are made to eradicate them. If cattle tracks, path edges or other resting places are not left to these species that once made up the full range of animal and plant life in the fields, then they disappear. This is all quite logical. Nobody should be astonished, least of all those involved in industrial agriculture, whose objective was and is to maximize profit. Society has sanctioned these developments with its financial support. Whoever is willing to pay money so that pesticides and slurry can be applied *en masse* will inevitably be confronted with this outcome.

The command economy in the agricultural sector devours billions and billions of public funds that could surely be put to better use. Unfortunately, this is by no means a new or novel finding. Politicians have been aware of this since the 1970s, but have not come up with any solutions. Variable subsidies are cobbled together, but when they lead to failure and additional costs it is the taxpayer who picks up the tab. This works because they cannot assign their tax liability to individual state expenditures: everything first goes into the pot and is then shared out. The public receives inexpensive food, although this has already been substantially pre-financed with tax money. Given that a third of it is thrown away, it must be far too cheap. For drinking water, we Germans pay double, since it was previously subsidized to enable the costly removal of pollution. The system has become parasitic, yet those affected do not notice because what is really happening is obfuscated so skilfully.

It was a stroke of luck for those who caused these various problems that climate change was discovered and 'exposed' as the alleged cause of all change. Since this discovery, there are no longer any guilty parties, as everything can be attributed to the climate or, if the change really has nothing to do with the weather, to global change. Climate change and global change are the two magic phrases that cover what should be kept hidden. For some reason, it is the nature conservation organizations that reach for this explanation most often in order to explain what is happening in nature. People seldom ask about the actual causes and the actors responsible anymore. Such an approach is cheap, yet hardly appropriate. It is even less acceptable to continue to fight against the past enemies, in a time-honoured manner, when this is fruitless. The evasion of real, soluble problems by labelling them as climate change reduces our sense of obligation to demonstrate the true causes and find effective measures here and now. The hoped-for results lie too far in the future. The best model cannot predict whether climate change will favour our moths and butterflies or destroy many of them, since its results depend on which underlying assumptions are used.

Nature conservation and nature enthusiasts

What is happening here and now does not need a model. The effects are clear and universally apparent. But nature conservation organizations in Germany would rather fight proxy wars against 'alien species' and do battle with their traditional enemies, such as road construction, over the construction projects that change the scenery, than use their power and their political weight against the principal agent of species extinction, industrial agriculture. Nature conservation laws are not directed against the perpetrators anyway. As paradoxical as this may sound, their effect is felt almost exclusively by nature enthusiasts to whom practically everything is forbidden. One might almost imagine that they were the main culprits of species extinction. An unbiased analysis of the German protection provisions would give the impression that nature enthusiasts, those who are most directly interested in animals and plants, ought to be kept away from nature so that they can cause no harm. It should be noted that the UK nature conservation laws are less draconian: the collecting of insects is not prohibited except in the case of a small number of species that are listed in the Wildlife and Countryside Act 1981, and for which a permit is required for close study. Yet in the decades since the German Nature Conservation Act has been in force and has been extended and supplemented by EU regulations, the protected moths and butterflies have not become more common; instead the 'list of threatened species' has become longer because the protection provisions do not offer the necessary protection. It is not because so many nature enthusiasts are chasing butterflies that their numbers are dwindling.

State nature conservation has failed to provide any proof that the danger to species emanates from people who are interested in moths, butterflies and other insects, including collectors. While birds have become very shy as a result of the intense pursuits that they have been subjected to, and often still are, and must therefore be observed through binoculars, most insects must be caught if they are to be studied more closely. Otherwise one is left with vague identifications such as 'bee'. No glimpse through a pair of close-focus binoculars will reveal which one of the hundreds of possible species of wild bees this is. (I often used a pair of these, which can be focused on a point just a metre away, to watch butterflies. A prerequisite for using them successfully is always that I already know the relevant species well enough.) It is easy to presume that collectors pose a risk. However, this is quite wrong, not only because it is (or was) the collectors and their collections that have provided us with information about the occurrence and abundance of moths and butterflies – particularly about conditions in earlier times, going back far into the nineteenth century. Without the private collections, the large collections in our museums would not exist. The nature conservation authorities would not have any basis for including species on their 'endangered list'. Yet for some reason the laws and regulations have been formulated in such a way that, ultimately, the collectors and observers are the only ones that are affected by them.

Whoever wishes to research wild bees and their occurrence in Germany and needs briefly to capture them in order to determine their species, requires a special permit under nature conservation legislation, since all wild bees are protected. If 'protected species' are killed through the incorrect application of pesticides by agriculture, we cannot even take them away to investigate the quantity of pesticide ingested without prior permission. If farmers wipe out whole colonies of wild bees along field and roadside verges, despite this being public land and not their own, or if water authorities carry out construction projects on dams, in the course of which tens of thousands of wild bee nests are destroyed, this has no consequence. Such interventions lie outside the effective scope of the species protection regulations. These examples are not trivial individual cases: anyone who spends time outdoors and is sufficiently observant will be able to give countless others. For example, the conversion of the embankments along the River Inn into an international long-distance cycle track destroyed huge colonies of sand bees and mining bees, without any interventions from nature conservation bodies. Many rare beetles, moths and butterflies were similarly affected. The wild bee losses were in the hundreds of thousands, if not the millions. Bumble bees, moths and butterflies have since fallen victim every year to maintenance measures because plants flowering along the embankments could apparently impair the safety of the cyclists that ride along them (see Photo 40).

The Devastating Effect of Communal Maintenance Measures

The situation regarding public maintenance in Germany is truly terrible. Maintenance measures along roads of every kind, on traffic islands and other communal areas, effect what are currently among the best habitats for moths, butterflies, wild bees and rare beetles, because mowing is carried out too soon, too often and at the wrong time of year. Sometimes, as was the case on a federal road in southeast Bavaria in 2018, the verges are mown as early as the end of April. At that time, the first flowers are just blooming by the side of the road. It is hard to understand how they are supposed to have obstructed the view of drivers or compromised traffic safety in some other way. Barely had the vegetation begun to recover when it was cut down again. Even worse than the hay-cutting in the meadows, this road-side mowing is often a 'close shave' that goes deeper than it should and tears up the top surface of the soil.

If deep-rooted plants, such as the beautiful chicory with its sky-blue flowers, nevertheless manage to bring forth new shoots and open their starry flowers again in July, this traffic hindrance must be removed immediately. Where there are flowers between the roadside and the maize fields, and where there is the slightest risk that they might obstruct sightlines, they are simply mown down. Evidently, agriculture enjoys full privileges, while flowers by the roadside have become unacceptable.

Perhaps it is nothing to do with traffic safety, but rather the use of expensive machinery, the procurement of which must be justified. After all, companies have been tasked with carrying out the maintenance measures if and when they happen to have the machinery available. But these places are not just odds and ends of land. If one were to add up the many million kilometres of the German road network, the many traffic islands, motorway verges and embankments and the other urban areas under the care of the public authorities, they would probably add up to an area greater than all our nature reserves combined.

It is difficult to believe that this unnecessary maintenance is carried out and paid for with public funds, that is, tax money, and that states and municipal authorities allow roadside verges and traffic islands to be used for the disposal of slurry. This massive application of fertilizer inevitably increases the intensity of the mowing thought to be required and effectively turns road traffic areas into agricultural spaces. This lunacy cannot be criticized strongly enough, since here it is not just industrial agriculture destroying insects. Naturally, the agricultural sector is interested in the returns from its fields, not in how many butterflies are flying over them. In terms of surface area, industrial agriculture takes first place by a long way when it comes to the destruction of species. But public maintenance takes second place. The state- and municipally sanctioned destruction of nature could be almost completely avoided, since no returns need to be obtained from their projects. True maintenance – for example, mowing to avoid sunny, open and insect-rich areas from becoming too overgrown – would only need to be carried out occasionally; for the sake of nature, not because machinery that happens to be free must be employed.

Industrial farmers now expect to receive additional payments from the public if they refrain from applying pesticides to a narrow strip of land beside their fields, so that a few wild plants can grow there and give the impression that there is still diversity in the fields. Given that the massive public subsidies to agriculture should, as a matter of course, include the obligation to farm the land properly, that is, without detriment or harm to nature, ground water or the air, it is quite absurd to destroy the flowers growing along the roadside next to these flowering strips. As so often, state and municipality do not provide a good example to the public but, instead, the worst example imaginable.

The same applies to the use of pesticides. Why may no blade of grass grow under a motorway bridge, and why is everything green around it discouraged with the application of herbicide? If a river flows under that bridge, as is so often the case, the poison will be washed into the water by the first heavy rain. One really does not have to be an expert in bridge statics and road traffic to question the necessity of poisoning the ground under motorway bridges. This is certainly not the only area in which unnecessary pesticides are applied, but if one wanted to briefly halt the growing vegetation, it would be enough just to mow the areas under motorway bridges once a year.

It is almost unbelievable that, even in state forests, the verges of the forestry tracks are regularly mowed. When this is done at the end of June or the beginning of July, it destroys the very plants that the caterpillars of many woodland butterflies live on, if they do not feed on trees and bushes. Moths, butterflies and bumble bees lose their essential sources of nectar. Rare, protected plants such as the Turk's cap or Martagon lily, *Lilium martagon*, are simply mown away. Surely no forester would be able to explain just what type of traffic safety such maintenance measures are supposed to serve. Flowers along the forest or woodland path certainly do not obstruct the gigantic timber harvesting machinery. And since the forestry tracks take such a battering, and because they are used all year round, they must be given a new surface every year. Often when this is done, all vegetation is pushed off the track edges by the bulldozers. The mowing of many kilometres of forestry track edge in early summer costs money; in this case, once again, public tax funds. The lack of flowers and butterflies along the forestry and woodland tracks reduces the possible enjoyment of nature in, and the recreational value of, the forests – both aspects that German state forestry is supposed to consider under forestry law (see Photo 41).

As has been suggested many times, private gardens should naturally be, or ultimately become, free of poisons. Given their combined surface area, they are the only habitats that can be truly meaningful for the survival of numerous moths and butterflies. These insects have virtually no chance of survival in the fields, now that these are subject to intensive industrial agriculture. They are not able to live along the roadsides and other communal areas, since typical German ideas of tidiness are clearly against it. The extent of the deficit in the general understanding of the natural world is evident from the fact that small areas that should be protected from mowing have to be provided with a sign announcing 'Flower meadow under cultivation'. Such places are fenced off with coloured ribbons. They must be wrested from the grip of municipal and state transport route maintenance, even though the regulations stipulate consideration of the interests of nature conservation. In our woods, too, the conditions for butterflies and moths are steadily deteriorating, albeit at a slower rate than in the fields. It is for this reason that practically the only representatives from the butterfly world we now see in much of Germany are cabbage whites and peacocks and a few migratory butterflies.

The End of the Night:
The Role of Light Pollution

A further objection could be raised concerning moths – namely, what about light pollution? Our times have brightened up the night. Millions and millions of lamps are alight from early evening until far into the night or even through to morning. This light attracts insects. Nobody can estimate even approximately how many moths fall victim to these nightly illuminations. One thing is certain, however, and that is that bats profit when insects are attracted by streetlights and collect there, as if for their benefit. Is it possible that the light even entices many insects away from the fields and into the brightly lit cities?

Assumptions of this kind do not seem unreasonable. Numerous investigations into insects that live in or near water have established that this effect exists. We have been familiar with it since the earliest use of electric light. Along river and lake shores, street lighting can attract aquatic and semi-aquatic insects that are swarming at dusk in such huge numbers that the street beneath the lamps becomes slippery with insect corpses, especially mayflies and caddis flies, but also large non-biting midges. Functioning on a smaller scale, insect-killing devices based on light attraction can currently be bought very cheaply for domestic use from discount supermarkets. This is happening at the very time when there is so much lamenting about the decline in insects. The devices do not seem to require a permit even though they kill protected as well as unprotected insects quite indiscriminately. Insect-killing devices are bestsellers and different discount stores offer them in various designs.

But to return to the question of what part so-called 'light pollution' could play in the decline of Lepidoptera. There is no doubt that the lighting on streets and buildings as a whole has a far greater effect than domestic insect electrocuting devices on terraces and balconies. By far the largest proportion of these lights have still not been replaced by a light mixture that has a less intense effect on moths than ordinary light, such as neon. Assuming that, for this very reason, comparative investigations into moths in the city would deliver no results, I only started my investigations using this method about a decade after having started work at the Bavarian State Collection of Zoology (ZSM) in Munich. The inner courtyard of the Nymphenburg Palace buildings, which shuts out light on all sides, seemed suitable to me, since it became dark in the evening and remained dark all night, according to the natural diurnal rhythm. The light from my window, which was, in any event, the only one illuminated on summer evenings, ought to have had no effect, since I had placed the moth trap lamp right next to it. Those were my thoughts at the time.

The insect flights that arrived were surprising in every respect. For the large moths, that is, everything that could not be classed as Microlepidoptera, which are notoriously hard to identify, I recorded the unexpectedly high number of 230 different species between 1981 and the beginning of 1985. The quantities recorded were also surprisingly high for a big city, or so I thought, even though they were significantly lower than those that still came to the light on the outskirts of a Bavarian village at that time. Three decades later, however, and they exceeded the quantities for 'the countryside'. In retrospect I deeply regret the fact that I did not continue with these investigations immediately following the relocation of the ZSM to the new buildings in Munich-Obermenzing. Again, I had assumed that any investigations would be unprofitable, since the new location was not protected from street and city lighting. The nearest streetlamps were only 20–30 metres away and, moreover, the light conditions were normal for a city. To my great astonishment, the insect catches turned out to be no worse at all. The quantities were at approximately the same level as in the unlit inner courtyard. However, the range of species was greatly increased: for the moths, as well as for the other nocturnal insects. My colleagues in the ZSM were very impressed by their diversity. I probably created quite a lot of work for them from time to time, with my frequent requests for help with identification.

As I have already mentioned, the 'super summer' of 2003 brought super results. One might have thought that the city lighting played no part; at least not as regards the UV light traps and the insects that responded to them. In the usual, time-honoured way, there were good nights and bad nights for results, corresponding to the weather. In late autumn and early spring, when the nights were long and naturally dark, the moths that were adapted to those times of year flew into the light as if there was no difference in the level of illumination. In the artificially lit city, there were in fact more common winter moths than in the countryside, and they represented the whole species spectrum that was normal for the region. The reaction to the moon was similar. Nights when the moon is full are generally unfavourable to moths, while nights with a new moon, in contrast, are favourable, even without cloud cover. Cloud and light rain are best; cloudless and windy conditions, on the other hand, very poor. All of this was substantiated with temperature measurements, which I noted down exactly for each night after I had investigated the insects that had flown to the light. It was not even possible to confirm more noticeable shifts in the flight times of the different species over the course of the year in comparison to those in the countryside.

Nevertheless, the findings relating to the influence of temperature are yet to be subjected to deeper analysis. In light of the discussions regarding the consequences of climate change and the related model calculations, such analysis could be very instructive, since the data goes back to 1969. The gap between 1994 and 2002 turned out to be unimportant. It is an example of how easily one can be influenced by opinions, especially those that seem plausible. Various short-term investigations that I carried out in the meantime at several places in the city ought to have alarmed me, but I held too fast to the notion that insect flight data were compromised and meaningless in the city due to its being permanently flooded with artificial light. However, the fact that many butterflies occurred on the site of the ZSM could not be ignored. I followed their development from the beginning, that is, straight after the move in 1985. It offered a useful lesson in many respects.

At the time of the relocation, there was a great deal of gravelly, raw soil resulting from the construction project. Exceptionally species-rich vegetation gradually grew over this. Butterflies arrived, among them several species of blue, and they rapidly developed impressive populations. At times, they were so numerous that a walk over the top of the subterranean construction would cause dozens of them to fly into the air at once. Yet it was evidently the opinion of the authorities responsible for the construction project that the ground should not remain in such a 'disorderly' state. The site was seeded with goat's-rue, *Galega officinalis*, a plant belonging to the papilionaceous plants or legumes. Symbiotic microbes on the roots called rhizobiaceae enable these plants to absorb atmospheric nitrogen.

This means that they grow very vigorously in poor soils provided enough water and mineral nutrients are present. Through their growth, they return nutrients to the soil in a very similar way to the clover fields that were once used for soil improvement in grain farming. In a few years, the goat's-rue grew to cover all the areas on the site where the soil surface was left open or torn by the construction works. At that point the blues and other butterflies disappeared. Their habitat had been destroyed through these attempts at 'greening'. The plans miscarried drastically, as the goat's-rue then had to be cut back for several years in order to control it. It grew chest-high as a dense, single species. It was thanks to the tireless dedication of the late Wolfgang Schacht, a colleague of mine at the ZSM, that about a decade after its mass growth it could finally be limited to some insignificant individual plants. Eventually, the butterfly populations also recovered. Mowing was kept to a minimum and never carried out over the whole area simultaneously, so that an area large enough for recolonization was always left available.

The ZSM became a centre of species diversity within the city with a richness in species that would have been worthy of a nature reserve. The numbers of insects recorded by means of light traps increased accordingly. Some areas of poor soil had loose, species-rich vegetation, while the rest were covered by a copse and diverse landscape structures provided by groups of trees, open land, hills, a pond and bushes as well as individual trees surrounding the parking areas: together this offered the structural requirements for a high level of species diversity in butterflies, moths and other insects. In places like this, man and nature can fit together in an encouraging way. The many species have opportunities to survive, at least a majority of them do, if their survival is permitted, just as in similar places in the city. The goat's-rue episode is a thing of the past and hopefully it will provide a salutary lesson: it is often the well-intentioned who take control and end up spoiling everything, since they are reluctant to simply let natural developments take their course, providing gentle steering if required. This is one of the key problems of moth and butterfly conservation.

Summary: A Cluster of Factors

Moths and butterflies have become more rare; far more rare than they were 50 years ago. In fields that are farmed intensively, they, like other insects, barely exist anymore. Their populations have been best preserved in gardens and parks in urban areas. Declines have taken place in forests and nature reserves. The extent of the decline is variable, as are the causes for it. The main cause, which has affected every area, is the overapplication of fertilizer. It changes living conditions fundamentally. Pesticides are certainly also partly responsible, although to what extent is controversial and may differ from region to region. The loss of structures in the landscape is more important than is generally thought. Maintenance measures applied to areas of land that are not used for agriculture affect many species of butterfly and moth. They are not necessary; certainly not with the degree of thoroughness and intensity with which they are carried out. The fact that the roadside verges are mown when flowers bloom there, even in state-managed woodland, is inexplicable. The reduced use of pesticides in gardens and parks will be beneficial for the moths and butterflies. Cities and larger villages have generally become save havens for them. However, the growing tendency towards low-maintenance, weed-free gardens is unsettling, especially if their goal is plant-free gravel or artificial plastic lawns.

Let us now look at the situation from an ecological perspective. The principal finding is that moths and butterflies have lost a great deal of their habitat. Around half of the total surface area of Germany has become inhospitable to them because it is covered with plants from which they cannot and should not live. Boundaries between fields used for different purposes have disappeared in many places as a result of land consolidations, and consequently the micro-climate, which has become colder and damper due to overfertilization, is less graduated. As a result, the generally warmer summers have, on balance, also offered nothing for the moths and butterflies, except in nontypical years, which do not have a lasting effect.

The vast majority of species that we are concerned with, if not all, can adapt well to the significant seasonal shifts caused by climate change, since they react to the same temperature signals as the plants. Moths and other insects also appear to cope better than was feared with the excess of light that influences their flying conditions in residential areas. Where they can live under semi-natural conditions, biological factors regulate their abundance. Parasites and probably also pathogens prevent excessive population increases, as they always have done. This is confirmed by the rarity of population explosions, since these usually collapse rapidly without having caused any damage. Where these do still occur, for example in commercially managed woodland or on fruit farms, it is due to the excessive uniformity of the tree populations. If they are genetically very similar or, as clones, practically identical, and the same age, as in single species plantations, they favour mass population growth. Control measures, particularly those based on pesticides, can achieve the opposite to their intended goal and delay the natural decrease of pests to insignificant numbers. This is because when pesticides are employed, many insects are affected in addition to those that are targeted, among them their natural enemies such as parasitic ichneumon wasps, Braconid wasps and tachinid flies.

The long-term cycles that occur with some insects, among them moths and butterflies, have still not been sufficiently researched. There are indeed indications of various causes, but what really triggers cycles of seven, ten or more years is as good as unknown. Quite sudden great surges in numbers of particular moths or butterflies can therefore be observed time and again, without this being a sign of their recovery. The same is of course true of the migratory butterflies and moths that arrive in their masses in certain years. Their vigour is immense where it is not supressed through pesticides. Whether pesticides are truly necessary should be questioned much more critically and discussed more openly from an agricultural-political perspective. There is no doubt that the failure to employ preventative pesticides can lead to a loss of profit. However, the use of pesticides also incurs costs and has an environmental impact. If one were to subtract these from the income achieved, it is quite conceivable that the pesticides would be shown not to be worthwhile and would only be necessary in exceptional cases, where mass population explosions of damage-causing insects were already taking place. Since the portion of public subsidies awarded to agriculture to pay for crop protection products is probably significant, the issues of economic sense and necessity are brought into sharp relief. Instead of granting subsidies upfront, a possible alternative would be to make them available in the event of substantial crop failures, as is the case with weather-related damage. The negative and far-reaching consequences for the natural world would thus be very significantly moderated.

The Disappearance of Moths and Butterflies and Its Consequences

Moth and butterfly numbers have been decreasing for decades. In the fields, which account for more than half of the land in both Germany and the United Kingdom, their current numbers represent perhaps a fifth of the abundance that existed prior to the major transformation in agriculture. My findings from southeast Bavaria are consistent with those of many others who have confirmed the decline.

The declines in the populations of Lepidoptera in Germany are paralleled over approximately the same period in Britain and North America. British butterfly species have declined by more than 70 per cent, 54 per cent of them contracting their areas of distribution, and the larger moth species decreasing by an overall 28 per cent.[*] In the southern half of Britain, the decline of the larger moth species has been much more serious, with a decrease of 40 per cent. Of the 337 common and widespread species of larger moths that were monitored between 1968 and 2007, two-thirds declined, 37 per cent of them by at least half. In addition to the 62 species of macro and micro Lepidoptera that became extinct in Britain during the twentieth century, a further 4 are now thought to have become extinct. At the same time, a third

---

[*] See Richard Fox et al., *The State of the UK's Butterflies* (Dorset, 2011); *The State of Britain's Larger Moths 2013* (Dorset 2013).

of the species have increased in number, 16 per cent of them by more than double over 40 years, apparently largely due to climate change. It is believed that, since 2000, a total of 27 species have colonized Britain for this reason.

In North America, the statistics over a similar period, although not quite as dramatic as in continental Europe, are nevertheless serious and worrying. They vary from state to state depending on where detailed observations or monitoring studies have been made. These reveal that, in general, southern species are increasing and extending their ranges northwards and northern species are decreasing and contracting their ranges even further north and to higher elevations, apparently because of the warming climate. An investigation carried out since 1972 at various natural and semi-natural sites in northern California by ecologist Dr Arthur Shapiro and his co-workers shows that butterflies are declining at all of them, even common species like the small cabbage white, *Pieris rapae*. Among individual species, the well-known monarch butterfly, *Danaus plexippus*, has decreased by 80 per cent since the mid-1990s, apparently at least partly due to the widespread planting of genetically engineered crops and to the use of a herbicide that effectively kills milkweed, the monarch's larval foodplant.*

Apart from climate change, the reductions in species appear to be due to the same causes that have been identified in Germany: loss and fragmentation of habitat and habitat degradation, urban and industrial development, unfavourable forest management, industrialized agriculture, overfertilization and overuse of pesticides and herbicides.

In Germany, as elsewhere, the regions with large-scale industrialized agriculture have been hit the hardest. In these regions, the nitrogen surplus per year is still around 100 kilogrammes of pure nitrogen per hectare or more. However, the whole country, indeed the whole of Europe, is overfertilized because quantities of nutrients are recorded in the air that correspond to earlier, whole-nutrient fertilization. Only very porous, sandy soils are less heavily subjected to this nutrient enrichment. It is least present in urban soils where tar surfaces and concrete allow the fertilizing precipitation to run off. From there it

---

* See also S. H. Black, 'North American butterflies: are once common species in trouble?' *News of the Lepidopterists Society* 58 (2016): 124–126.

flows through the sewer system into sewage treatment plants, although where the soil is very porous it may also flow into the groundwater. Forests and woodlands retain the most and can exploit a large proportion of the fertilizer from the air directly. In lakes, especially in areas that become very warm in summer, the nutrient supply encourages so-called 'algal blooms'.

Overfertilization is thus a general problem that is hugely aggravated by the extremely high use of slurry and artificial fertilizer in agriculture. Its signature, left behind in sediments and other residues, will characterize our period in the distant future, just like the impact of large meteorites that leave their marks on our planet. For decades, ecologists have tried to draw attention to this phenomenon. The Ecological Society of America published a special brochure in 1997 on the human alteration of the global nitrogen cycle: its causes and consequences.

We are already paying very dearly for one of the main consequences in our provision of perfect drinking water. The second major impact of industrialized agriculture – that is, the contamination of soils and the residue of contamination in foodstuffs – is once again being fiercely debated. The main issue under discussion is whether the substances in question, which are verifiable in quite small quantities, are carcinogenic or damaging to health in some other way, like DDT. This substance has ultimately been made broadly illegal following heated disputes because, among other things, its secondary effects harmed many animal species that were not the target of its application. In *Silent Spring*, published in 1962, the American marine biologist and environmental activist Rachel Carson directed attention to this type of collateral damage, which had been initially ignored and suppressed until the public took notice. One could not allow a 'silent spring' without birdsong to come to pass: this was the unanimous opinion of nature and environmental conservationists. They prevailed, and the battle appeared to have been won.

DDT was banned but not completely relinquished, and it continued to be used in the tropics. In North America and Europe, persistent insecticides with their long-term toxic effects were replaced by pesticides and herbicides that decompose (more rapidly) and are more accurate. But these too have turned out not to be the hoped-for panacea. One thing after another has had to be banned, and this continues to the present day. Unfortunately, this focuses attention on insecticides and thus away from overfertilization. The consequences of mechanical interference in the form of maintenance measures have also been largely overlooked. Accordingly, these have particularly affected the moths and butterflies in the marginal areas that are not processed agriculturally. Protection and recovery for animal and plant species should actually be provided by the open landscape. However, it was only when such steep declines in insect populations were recorded in nature reserves that they made the headlines in the second half of 2017 with publication of the Krefeld Study that the conservation community was reminded of the marginal strips bordering agricultural land and other communal areas. The methods for processing these areas have long since become entrenched, because it is hard to change something once it has become routine.

The decline of insects led to many questions being asked by members of the general public. What do we actually need these tedious creepy-crawlies for, in particular the moths? We all know that bees are useful and necessary, but other insects? There are enough butterflies flying around in the garden. Most of them are moths and thus, like many insects, vermin. We should be happy, they say, not to be as plagued by mosquitos as in the past, or by wasps and other things that are just annoying or that sting us. Only the death of bees worries agriculture, since oilseed rape and fruit trees are dependent on them for pollination. We already have prototypes for technical mini drones to use as substitutes for the real thing, although they are not ready for series production. Therefore, to bridge the gap, the agriculturalists argue, we still need to maintain bees in the necessary quantities, but certainly not flies. In earlier times, they annoyed the cattle both in their stalls and outside in the meadows. Now that manure and water are taken off as slurry, the dung-heaps where their maggots developed have been removed. Improved hygiene in modern cattle stalls ensures that there is nothing for flies to find.

Apart from that, the insects that used to exist in fields and meadows were damaging to crops, weren't they? Grasses do not need insects for pollination because the wind does it for them. Even the wind is not needed, it is asserted, since most grasses should not even be allowed to flower in the meadows. Meadows used to be mown several times a year. One fights grass flies and stem sawflies in order to prevent them from damaging the crops, and also many beetles, bugs and, of course, aphids. In fact, a 96 per cent reduction in insect abundance, as was demonstrated by the light traps in the fields, was exactly what was meant to be achieved from the perspective of agriculture. The fact, they conclude, that the disappearance of 70–80 per cent of these insects since the 1990s has been established in the nature reserves of northwest Germany need not be grounds for panic, need it?

Furthermore, it is argued, most species of insect fall within the category of pests, and many butterflies too. There are still too many cabbage whites in our gardens. This is evident enough from the damage caused to cabbages and other plants. Codling moths and common winter moths are among the true pests in horticulture and fruit production. So it is great that they have become rarer, isn't it? We can do without masses of ermine moths, too; not to mention oak eggars or pine-tree lappets, whose caterpillars have hairs that can trigger violent allergic reactions. This is a rough summary of the reactions with which I am confronted at lectures and discussions on the subject of insect extinction.

So, let us not be under any illusion. The position of insects in public opinion is not all that good. The fact that they are interesting and that there are some beautiful species, especially among the moths and butterflies, which are moreover quite harmless, is not quite enough to ensure their survival. Most people see everything from the perspective of utility. Butterflies and moths should therefore provide at least an 'important contribution' to the pollination of crop flowers in order to justify their existence. Very few of them do. They did not evolve in order to do something for people. They live their lives and use the opportunities that are available to them. The question of whether this is actually 'useful' for us humans, as in a few cases it is, constitutes an evaluation, not a fact of nature, nor even a stipulation for nature. Even the personification of 'nature' is an expression of human thought, since 'nature' as a discrete entity does not exist. We need to construct an ecosystem intellectually if we want to alter it, since it is not prefabricated. Natural processes run the way they do, because they do, and not because they ought to do so; there is no superior authority, although some people would gladly like to believe in one. Western philosophy provides food for such arrogance. We should bear this in mind when we pose questions about the necessity of creatures such as moths and butterflies or attempt to provide answers about securing biodiversity for the future.

What sort of answers can we give? There is a common one that is easy to explain: butterflies, moths and other insects are significant because birds depend on them for food. The truth of this statement is borne out by the parallel decline in species of field birds. Across Europe, their populations have declined by more than half since the 1990s. In Germany, the loss in areas that are heavily used for agriculture constitutes over 90 per cent. There are hardly any skylarks left over the fields, yellowhammers at their edges, or partridges and quails in the meadows. Grassland birds have been particularly hard hit because grassland is mown far too early in the year and too often. Concerning the decline of the field birds, the 'silent spring' that Rachel Carson warned of has long been true. It should now be called 'silence across the fields'.

Whoever wants to hear a chorus of birdsong in spring would do best to go to a city park or a city cemetery, since birdsong and bird calls are disappearing from commercially managed woodland too, and the sounds that were once so varied are becoming monotonous. Caterpillars and other insects provide nutrition for birds. If these insect populations decline, then bird populations will inevitably shrink. The fact that they manage to maintain high population levels in the big cities despite all the obstacles that doubtless exist there says less about the quality of the cities than about how miserable living conditions in the fields must have become. The example of the skylark serves to illustrate how paradoxical these conditions can be. The best chance of seeing singing larks ascending into the air in southern Bavaria is provided at Munich airport. There is barely anywhere of comparable surface area far and wide that has as many skylarks. Conditions are, of course, not quite so good for enjoying their song.

So far, so unsatisfactory. Naturally there are some who will pose the follow-up question: why do we need larks and other birds at all? For example, swallows, whose numbers have declined very sharply. If there are too few insects, then we do not need the birds to keep them in check anymore. There will not be so many migratory birds captured around the Mediterranean if fewer and fewer of them fly along those routes; the decline in migratory birds will perhaps put an end to the hideous practice of capturing songbirds. This type of logic is incontestable if it starts with the question of utility. Anyone who has ever planted something in the garden that they wanted to use and then watched it grow, whether flower or vegetable, can relate to this. Even if our attitude starts to change over time, we must admit that gardens are not usually created for insects and birds, but to serve our personal needs and expectations. It takes effort to allow anything that turns up in the garden to stay there, and whatever is good for insects, to grow; perhaps even a readiness to argue with one's neighbours. Gardens should not be allowed to run wild, and this is still a valid view. It is probably the reason why there was barely any reaction, let alone formal resistance, to the maintenance measures employed on embankments, roadside verges and public open spaces, even if these are not used for recreational purposes and are therefore easiest to maintain if they are close-cropped.

This utilitarian thinking pervades all areas of life. It must be consciously qualified, better still suspended, if 'nature' is to be allowed to unfold by itself, at least in certain areas. This achievement is what we call 'culture'. We are not just willing to pay for it, but willing to pay a considerable amount. Through our participation in cultural life and its promotion, we show ourselves to be cultured, since culture is something that goes beyond what is purely utilitarian and necessary. Hubert Markl, former president of the German Research Foundation [Deutsche Forschungsgemeinschaft] and the Max Planck Society, thus considers appreciating nature as a cultural activity.

The decisive answer to the seldom well-intentioned question, 'Why do we need moths and butterflies?', stems from this idea. Whoever genuinely asks this question deserves to be asked in turn: 'Why do we need an agricultural sector that is so highly industrialized?' It has long since surpassed the reasonable production of necessary foodstuffs of the desired quality. The evidence for this harsh observation is the fact that twice as many organic products are purchased in Germany as are domestically produced. The demand is there, yet only half is covered by domestic production. German agriculture produces food on a massive scale for the global market. In recent decades, there has been permanent overproduction. In order to reduce the surplus, the general public is forced to use petrol with added biodiesel. Production quantities were targeted towards the maximization of profit, not the demands of the people who paid the subsidies. At the beginning of the twenty-first century, even grain was burnt as biomass, since the quantities produced had become much too large, the price for organic grain had sunk, and thus proceeds were greater if it was sold as fuel. Farm-based agriculture, which had formerly been diverse and designed around local demand, became a victim of global competition, spearheaded by large industrialized companies. Whether this demonstrated a responsible attitude on the part of the state may well be questioned, although this is exactly what subsidies ought to promote.

We can also look beyond the fields and meadows. Do we need artists, concerts, historic preservation or even science, if we are only concerned with utility? In a world geared only towards profit, these are luxurious and dispensable, just like the songs of the birds and the sheen on the wings of a butterfly, or the poems by authors who have admired butterflies, such as Hermann Hesse. The profit-oriented economy does not need these either – writers, who convey the pleasure and stimulation of reading, relaxation or creative tension. We must steel ourselves against such attitudes; all of us. Profiteers cannot be allowed to prescribe what kind of animals and plants may survive or which creatures should give us pleasure. The virtual world created by screens is no replacement. Quite the opposite: we should withdraw subsidies from all those who directly or indirectly encourage the sacrifice of the natural world in our fields. They should live by our will, not the other way around, even if the current system obfuscates this. Industrialized agriculture is far from being the functioning symbiosis that we need, because the system will only last if both partners extract the same level of use from it. If one side uses the other side too much, it becomes a parasite. This tendency is unquestionably present, but we do not need to accept it any longer. This, dear readers, if you will, is the message of the moths and the butterflies.

We can justify putting a high value on their existence, their beauty and their unique lifestyle: one that politicians must take into account. Our reluctance to allow moths and butterflies to disappear, birds to die out and flowers to stop flowering need not be based on the primitive concept of utility.

What We Can Do about the Disappearance of Moths and Butterflies

Making industrialized agriculture environmentally compatible in the foreseeable future is an objective that must be pursued, but not an option that is likely to be successful in the short term. It will continue to hold its course like an overloaded super-tanker, whatever it may cost us. We are paying the fare, after all. Many farmers will give up their businesses, forced to do so by the overall prevailing conditions, just like thousands of other farmers had to over the past decades. As a sceptic, one must assume that a system as complex as that of EU agriculture must collapse before it can be reformed. Quite how such an implosion plays out was demonstrated by the collapse of the Soviet Union in 1990. As an optimist, one must be steadfast in one's faith in the strength of parliamentary democracy and hope for reforms that can bring about real change, on the principle that 'hope dies last'.* The situation should be significantly easier in the municipal and state sector in Germany, especially since here the issue is not one of profit, and, indeed, considerable costs could be saved if excessive and superfluous maintenance were to stop. However, experience shows that even these systems are not as easy to reform as one might think. Much of current practice has become habitual and is therefore entrenched. Anyone who

---

* This is a Russian proverb: *Nadezhda umiraet posledney*.

has assumed, up to now, that they have done a good job will resist any changes that put that work in a bad light – all the more so if such alternatives run counter to the German obsession with order and cleanliness. Leaving disorder behind, permitting uncontrolled growth, even if just for a short while, goes against German nature. At least, that is the cliché.

It is more difficult to challenge and attempt to change laws and regulations that have been valid for many years, since this threatens the principle of legal certainty. 'In the past it was like this, and things were fine', some people will say. 'Now it is supposed to be different, and this cannot be allowed to happen.' This 'resistance logic' is deeply anchored in our psyche. It is not legal sophistry. Demands will be made of lawyers, if they find it difficult to accept nature and its diversity as it actually is. For reasons that are essentially understandable, they want to tidy everything away in clearly labelled boxes so that there are no grey areas.

Even large nature conservation organizations behave in accordance with this principle, never letting go of something that has been attained, never mind how meaningful and sorely needed a change to this position may be. They prefer to let their activists act in the grey areas because it is simply impossible to act in accordance with species protection law, to protect, in advance, everything that serves nature conservation and increases knowledge about animals and plants, since one cannot know what the future will hold. The shredded butterfly of a protected species, stuck in the radiator grill of a car, remains protected in Germany, even if it is dead, and cannot even be removed by hand without authorization. Sewage treatment facilities are exempt from such laws, just like the poison that kills protected butterflies, if it is legally applied by agriculture or in the garden.

But what should one do if a purple emperor flies onto one's hand or a common blue lands on one's chest while one is sun-bathing (see Photo 42)? The endless string of absurdities that the nature conservation authorities would have to deal with under the species protection provisions will not be unravelled here. It is clear what is at issue though. Species protection legislation and regulations should be simplified and adapted to reality. Their chief guiding principle should be that only those prohibitions or orders that are demonstrably necessary and contribute something by way of protection should be included. Such a rational solution would massively reduce the burden borne by the authorities, who surely have more important things to do, and would provide far better access to nature and its diversity to all interested parties, from children to amateur researchers and university research departments. Under current conditions, one must conclude that the extremely restrictive nature conservation laws only serve those who are responsible for the losses to nature, since they prevent the critical public from investigating their actions.

What position can realists take in this conflicted situation? I am convinced that the path must be from the bottom upwards, from the foundations to the top of the organizations, authorities and political committees. The objective must be for the critical public to become more interested in the species. We should concentrate on the beauty, individuality and unique characteristics of butterflies, moths, beetles, wild bees and other insects, as well as our wildflowers, rather than speaking of imminent doom because one species or another is threatened with extinction. The approaches taken by local nature conservation societies and many specialist amateur researchers cannot be praised highly enough in this regard. It is only if children and young people, who can still be inspired, and adults, who feel a sense of responsibility for the conservation of biodiversity in our land, admire moths and butterflies at first hand, follow the lives of their caterpillars and other insects and get to know the whole animal and plant kingdoms, that these creatures will attain the cultural status that they deserve.

Inspiration does not grow from afar but through proximity. The caterpillar must be allowed to crawl over one's hand, the large (highly protected) hawk-moth be allowed to sit on one's fingertip or chest. We must explain to our children why there are no more moths and butterflies flying over the fields. They ought to know why no more sky-larks can be heard singing. They should not require a permit to enjoy nature or be locked out of it in those very places where they can still experience diversity in animals and plants. Children should be allowed to play outside again. This will require relatively natural urban spaces that have not yet been built on and also access to the countryside, paths through meadows and fields and, of course, also through forests. 'No entry' is the worst kind of protective measure. It should only be used where it is strictly necessary in order to protect sensitive species from disturbance. There it should apply to all people equally and without exception. We should no longer acquiesce to the blocking of access to nature reserves to those who are interested in nature, while hundreds of anglers are freely entitled to trample down their spots in the reeds along the riverbanks or rid them of plants using glyphosate and even travel over the water surfaces with their boats, regardless of whether this disturbs protected bird species or not. It is equally unacceptable for hunters to put down food to attract wild boars to clearings in nature reserves where there are rare butterflies. The wild boars churn up the ground and change the vegetation, which is also influenced by the bait. Hunters and wild boar are free to roam in protected areas in Germany; nature lovers are not. Opposition to this must be shown from below, by the local people, under the leadership of knowledgeable environmentalists. Constructive cooperation with hunters and anglers is more likely to come about if they are confronted by people from their local area than by strangers.

The same applies to the local nature conservation authorities. They bear the brunt of what takes place on the ground and are familiar with the problems specific to the area. Therefore, they will surely also be prepared to bring about agreement between landscape conservation organizations, road authorities and community care teams, so as to achieve meaningful procedures regarding the care and maintenance of public spaces. Those responsible for state forests will not be able to escape such initiatives, but will participate in their own interest, in order to maintain and improve habitats and refrain from unnecessary destruction, such as the mowing of flowers at the verges of forestry tracks in early summer (see Photo 43). Activities such as 'A biotope for every community' [*Jeder Gemeinde ihr Biotop*], led by Peter Berthold, former director of the Max Planck ornithological institute at Radolfzell, provide sufficient evidence that this route up from below functions much better than that other, more customary one involving new laws and regulations imposed from above.

For nature conservation organizations, this means investing as much as possible in the acquisition of land and the treatment of such land in a manner consistent with conservation aims, and leaving aside any 'actions' that are clearly ineffective. The world will not be saved by apocalyptic prophecies, but through good examples and local behaviour. 'A biotope for each community' could be nicely extended to 'A butterfly meadow for every municipality' so that children can once again experience contact with moths and butterflies at close range, as well as grasshoppers, crickets and other creepy-crawlies. There are plenty of public spaces that could be used for this. Nature conservation organizations could also lease agrarian land, for example for five or ten years, and use it to demonstrate all the things that can live on our land when it is not poisoned or fertilized to death. The earlier 'set-aside land' that was supported with EU agricultural funds had a similar objective, but its effect was limited because the land was either left unused for too short a period of time, or left in such a state as to guarantee an immediate resumption of control. A meadow leased for five years does not need to deliver high yields like high-performance pastureland, although it will yield enough green fodder or hay to finance an annual mowing at the right time. It is best for the wildlife in the meadow, and will not disadvantage later uses of the land, if it is not flooded with slurry at that point. And so forth.

There are many options for taking action at the level of the community and the private landowner. Exactly what and how much is achieved may be taken as an indicator of the seriousness of the efforts of the nature conservationists. I know from my own home district and neighbouring districts that it is a great deal. District-owned spaces are in a better condition than state-owned nature reserves. Therefore, it would not do to be pessimistic: there are realistic chances and opportunities; anyone who is active on behalf of the natural environment knows how hard they are to implement, yet the dedication invested is worthwhile. If our children and grandchildren can still realize this in years to come, then it will truly have been worth the effort. This, too, is the message brought by the moths and the butterflies.

The Beauty of Moths and Butterflies

The word 'butterfly' is too plain, too prosaic to describe what makes these creatures so distinct. If we watch a peacock butterfly as it drinks nectar from the flowers of a butterfly bush in a park or garden, the fragile beauty of a creature such as this becomes evident. Wings as fine as tissue paper, with patterns and colours that an artist could barely imagine without the living model; antennae that pick up signals from the environment and convey them to the butterfly; eyes made up of many tiny ommatidia that register movement much more quickly than our own far larger eyes; and legs with a tactile sensitivity far more acute than our fingertips – yet all this only produces a rough, superficial impression of the essence of a butterfly (or moth).

Others that tumble about the flower heads next to the peacock, such as the small tortoiseshell, the red admiral, the painted lady and the large white, to name just the most well-known species, are similar in their kind as Lepidoptera, but essentially different in the colours and patterns of their wings. Whoever glimpses the beauty of butterflies at the right moment will be captivated by them. It is so multifarious that not even the largest museums have specimens of every species, even if their stock consists of more than 10 million specimens, like the Natural History Museum in London, the largest collection of moths and butterflies in the world. More than 260,000 different species are now known, have been described and given a unique scientific name. There are probably thousands more that are still unknown and awaiting discovery.

Countless collectors are committed to their research, most of them fascinated by the beauty and diversity of moths and butterflies. The variations and aberrations that occur in nature move some people to think nothing of the time and money required to discover them and incorporate them into their collections. Variations are the material from which evolution creates something new and increases biodiversity. The joy of discovery and aesthetic pleasure come together in the collection of Lepidoptera and can become addictive – an addiction that can seem like an illness, which causes otherwise highly respectable people, especially men, to behave in ways that seem disconcerting and even ridiculous. The painting by nineteenth-century German Romantic artist Carl Spitzweg, *The Butterfly Hunter*, depicts one such collector who can invest a great deal of money in his passion.

Museum collections have had, and continue to have, trouble with collectors who are driven half-mad by greed. The objects of their desire, the rare beauties of the butterfly kingdom, among them those that the inexperienced glance would dismiss as mere 'moths', achieve high prices on collectors' markets. However, international regulations designed to protect endangered species have created legal 'grey areas', blurring the lines between what can and what cannot be legally sold, and such markets have become associated with illegal activities. As a result, most private collections end up in public museums, and collections can seldom be passed from generation to generation, even among dynasties of collectors such as the Rothschilds. Often, private collections will come 'to the museum' after the death of the collector. There is an increasing tendency, owing to the conditions of our times, for agreements to be entered with museums during the lifetime of the collector. An example of this is the Thomas Witt Collection, which, at the end of the twentieth century, was arguably the largest private butterfly and moth collection in the world. It was assigned by its owner to the Munich Zoological Collection, together with a researcher to look after it. This is how one of the largest butterfly and moth collections currently in existence came about.

The passion for collecting is widespread; it includes almost everything that can be collected. Collections are not always distinguished by the aesthetic of their objects. These can be as banal as bottle tops or as valuable as coins. However, with butterfly collectors, beauty is almost always a primary factor and collections are designed accordingly to the extent that financial circumstances permit. Occasionally, they probably become overextended. The distinction between the beautiful collection that is enjoyed again and again in a small, still room or appreciated with like-minded friends, and public displays or the increasingly popular 'butterfly houses' in which one can enjoy living butterflies, is fluid. It is impossible to establish boundaries that would separate the scientific or scientifically valuable collections from the purely aesthetic. Neither would it be right to do so, since what is 'of value' in any area depends on the spirit of the age.

Who would guess that Winston Churchill would find moments of recuperation from the dreadful events of the time, admiring his butterfly collection while England was attacked by German bombers during the Second World War? Ernst Jünger, a German soldier in the First World War, described with incomparable skill how, during the 'storm of steel' [*Stahlgewitter*], he collected beetles in the trenches and found joy in their subtle beauty. During the collapse of Tsarist rule in Russia and his subsequent escape, Vladimir Nabokov drew strength from the butterflies that he had collected since childhood and that lived on in his literary works. Far more well-known and published in numerous editions are the 'butterflies' of Hermann Hesse. He dedicated poems of timeless elegance to them, like this one entitled 'Blue Butterfly':

A small blue flies by
Moved by the wind
A mother-of-pearl shiver
Glitters, flickers, vanishes.

Thus, in the blink of an eye
Thus, as it drifted past
I saw fortune wave to me
Glitter, flicker, vanish.

Whether it is the beauty of butterflies or of beetles, it inspires the eloquent to poetry and prose that preserves those fleeting moments when the wings flash open. The ancients felt this without having any deep zoological knowledge. The Greek word 'psyche' means both 'butterfly' and 'soul'. At the moment of death, the soul leaves the body like a butterfly, hurrying away from our gaze and disappearing. The ancients were familiar with the 'complete' metamorphosis of Lepidoptera and puzzled over it. The caterpillar, so different from the butterfly, eats and eats until it seems close to bursting. It moults repeatedly, a new caterpillar wriggling out of its skin every time. But then, in the final moult, a mysterious pupa emerges with none of the features that distinguished the caterpillar. It has no mouth, no anus, no legs, nor even any wings, unlike the butterfly that will emerge from it. It is packed into a rigid shell like a mummy. Only the lower section can make small twitching movements. At best, wings are indicated on the outside, as if pre-modelled by the finest carver.

Then suddenly, at a moment nobody can divine, the pupa bursts and the miracle takes place: the butterfly emerges. The miraculous pupa is known as a chrysalis and it is inside this that the transformation into a butterfly, which is still not fully understood, takes place. We know that it is controlled genetically and by hormones, but what a change of form it is. The worm-shaped caterpillar turns into a butterfly that bears no resemblance to its origins. It is hardly surprising that scientists have racked their brains over how this transformation, this metamorphosis, could have come about.

A new theory, submitted for discussion a few years ago, relies on what can be seen when caterpillars are parasitized. On page 54, I described how the large white caterpillars crawled up the wall of the house, gave up, remained hanging and, after a while, became covered in small yellow cocoons, inside which parasitic Braconid wasps were developing. According to this theory, one should imagine the development of the complete transformation in a similar way. In prehistoric times, worm-like articulates were attacked by parasites that developed inside their bodies and, after a time, emerged from the empty, devoured hull. Just as it still happens today. However, over time, according to the interpretation suggested by this new theory, the parasites planted their own genetic material in the 'worms'. This became combined with that of the latter and the genes have since then enabled two development programmes to proceed consecutively; first, what is now known as the caterpillar stage, and, second, that of the butterfly. In the interim, almost everything remaining in the body of the caterpillar is 'melted down' and a new beginning in terms of differentiation and development takes place.

This is apparently the secret of metamorphosis. From the original parasitization, the new, quite different form emerges, a symbiosis of both genomes. What would writers like Hermann Hesse have made of this? For them, the butterflies were already sources of wonder. Even more wondrous things are supposed to have occurred in their genesis in earliest prehistory. Whether they are correct, or simply charming reflections, such theories also demonstrate how metamorphosis continues to fascinate even rational evolutionary biologists. Butterflies are miracles. We should not let them disappear from our world.

Two Findings in Place of an Epilogue

Flying over the land in which we live, with good visibility – so that's how it is down below! Right-angled, straight-edged, no blade or bush that did not first have to prove its utility. Otherwise those aptly named plant protection products will be applied, and that will be that!

There was a time when we still had bushes and green hiding places and not this neatly divided European maize-production zone, cleansed of nature, that is called Germany – at that time we still heard the flapping of the bulbuls' relatives, our local nightingales.

Quoted from 'Highlights' of the *Süddeutsche Zeitung*
31 March 1993 and 2 November 1994
Josef H. Reichholf
Late June 2018

Pale clouded yellow

The yellow butterfly hurries away
through the bright pine grove
The sweat melts on one's brow
And summer once again gathers

all its strength
Becomes hotter than ever
so that Nature can be passed on
to the morning

Then we will be ready for contemplation
since we cannot hold onto anything

He searches on for his female
the small male pale clouded yellow
however short or long it may be,
his life.

Miki Sakamoto
'Goldene Acht', from *Vergängliche Spuren* [Fleeting Traces],
Kessel Verlag 2014

Select Bibliography

I have put together here a selection from the wealth of publications available that are either directly related to the topics and research results dealt with in this book or that are recommended for further reading. I should stress that this represents a personal selection. For descriptions of regional conditions, there are a great many publications available that, for the most part, have appeared in the periodicals of entomological or general natural history study groups or societies. Many of these societies are connected to or directly based at research museums. Of particular importance in Germany are the surveys carried out and coordinated in recent times by the UFZ-Umweltforschungszentrum (Centre for Environmental Research) Leipzig-Halle GmbH, in which amateurs and 'citizen scientists' took part. Anybody who is interested in becoming involved is encouraged to get in touch with the UFZ (the Helmholtz Centre for Environmental Research, www.ufz.de). The same applies to the federal state museums of Austria and Switzerland, which also coordinated investigations. In Britain, one should in the first place contact Butterfly Conservation, Manor Yard, Shaggs, East Lulworth, Wareham, Dorset BH20 5QP (email: info@butterfly-conservation. org or see www.butterfly-conservation.org).

Asher, J. et al., *The Millennium Atlas of Butterflies in Britain and Ireland.* Oxford University Press, Oxford 2001. (A comprehensive distribution atlas with details of the varying fortunes of the British species.)

Barkham, P., *The Butterfly Isles. A Summer in Search of Our Emperors and Admirals.* Franta Publications, London 2010. (Watching butterflies is very popular in the United Kingdom. Delightful descriptions by a knowledgeable enthusiast, which show that it is possible to write as beautifully about butterflies as it is about birds.)

Beebee, T., *Climate Change and British Wildlife.* Bloomsbury Wildlife, London 2018. (A readable nontechnical up-to-date survey.)

Bellmann, H., *Der Kosmos Schmetterlingsführer.* Kosmos, Stuttgart 2016. (Selection of more common and striking species with very good photos including of caterpillars.)

Black, S. H., 'North American butterflies: are once common species in trouble?' *News of the Lepidopterists Society* 58 (2016): 124–126.

Burton, J. F., 'The responses of European insects to climate change.' *British Wildlife* 12 (2001): 188–198.

Burton, J. F., 'The apparent influence of climatic changes on recent changes of range by European insects (Lepidoptera, Orthoptera).' *Proceedings of the 13th International Colloquium EIS, September 2001* (2003): 13–21.

Carson, Rachel, *Silent Spring.* Penguin Modern Classics, London 2000 [1962]. (The author's prediction has come true in the fields.)

Collins, N.M., *Legislation to Conserve Insects in Europe.* Amateur Entomologist's Society Pamphlet No. 13, 1987. (Sets out in detail, where available, the legislation of most European nations, including Germany and the United Kingdom.)

Dennis, R.L.H., *The British Butterflies. Their Origin and Establishment.* E.W. Classey, Faringdon 1977. (Exemplary analysis of how the world of butterflies has developed since the last Ice Age.)

Dennis. R.L.H., *Butterflies and Climate Change.* Manchester University Press, Manchester 1993. (A comprehensive but somewhat technical account concerning European butterflies.)

Dierl, Wolfgang and Josef H. Reichholf, 'Die Flügelreduktion bei Schmetterlingen als Anpassungsstrategie.' *Spixiana* I (1977): 27–40.

Dohrn, Susanne, *Das Ende der Natur. Die Landwirtschaft und das stille Sterben vor unserer Haustür.* Herder Verlag GmbH, Berlin 2018.

(Stirring, excellently written book about the effects of industrialized agriculture; focus on northern Germany.)

Ebert, Günter (ed.), *Die Schmetterlinge Baden-Württembergs*, vols. 1–9. Ulmer, Stuttgart 1991–2003. (*The* handbook for central European Lepidoptera and their biology.)

Fortey, Richard, *The Wood for the Trees. The Long View of Nature from a Small Wood*. Knopf, London 2016. (Nature writing at its best: makes clear how much living conditions in nature are changing in our times.)

Fox, Richard, 'Butterflies and moths', in D.J.L. Hawksworth (ed.), *The Changing Wildlife of Great Britain and Ireland*. Taylor & Francis, London 2001, pp. 300–327.

Fox, Richard, et al., *The State of Britain's Larger Moths*. Butterfly Conservation and Rothamsted Research, Wareham, Dorset 2006. (First comprehensive study to establish the decline of butterflies and moths – a decline by a third in the UK between 1968 and 2002. The declines described here and the findings from south-east Bavaria set out in this book are almost frighteningly similar.)

Fox, Richard et al., *The State of Britain's Larger Moths 2013*. Butterfly Conservation and Rothamsted Research, Wareham, Dorset 2013.

Fox, Richard et al., *The State of the UK's Butterflies 2011*. Butterfly Conservation and the Centre for Ecology & Hydrology, Wareham, Dorset 2011.

Habel, Jan Christian et al., 'Butterfly community shifts over two centuries.' *Conservation Biology* 30 (2015): 754–762. (Extremely important research that substantiates how large our species losses have been since the nineteenth century, even in a nature reserve near Regensburg that does not have any intensive agriculture nearby. The long-distance effects are far more extreme than had been assumed.)

Haddad, N., *The Last Butterflies. A Scientist's Quest to Save a Rare and Vanishing Creature*. Princeton University Press, Princeton 2019.

Hallmann, Caspar E. et al., 'More than 75 percent decline over 27 years in total flying insect biomass in protected areas.' *PLOS One* (https://doi.org/10.1371/journal.pone.0185809), 18 October 2017. (Includes the key findings of the so-called Krefeld Study that established that the quantity, the biomass, of insects in north-western German nature reserves had decreased by more than 75 per cent since 1989.)

Haslberger, Alfred and Andreas H. Segerer, 'Systematische, revidierte und kommentierte Checkliste der Schmetterlinge Bayerns (*Insecta*: Lepidoptera).' *Mitteilungen der Münchner Entomologischen Gesellschaft* 106, Supplement (2016). (Updated and complete overview of all Lepidoptera species documented in Bavaria, with commentary; in total, 3,243 species.)

Herschel, Kurt, *Falter bei Tag und bei Nacht. Aus dem Leben unserer Schmetterlinge.* Neumann, Berlin 1953. (For a change, some 'historic' descriptions that are refreshing to read.)

Hesse, Hermann, *Schmetterlinge.* Compiled by Volker Michels. Frankfurt 2011. (Poems and short essays about butterflies and moths; Hermann Hesse was an enthusiastic butterfly collector.)

Huemer, Pert and Gerhard Tarmann, *Die Schmetterlinge Österreichs (Lepidoptera).* Museum Ferdinandeum, Supplement vol. 5. Innsbruck 1993.

Kaltenbach, Thomas and Peter Victor Küppers, *Kleinschmetterlinge beobachten – bestimmen.* Neumann-Neudamm, Melsungen 1987. (As a field guide, this offers an appropriate selection from the abundance of Microlepidoptera that can often only be identified by specialists, with brief descriptions of their lifestyles.)

Karl, Gerhard, *Kleinschmetterlinge in Südostbayern.* BUND Naturschutz in Bayern, Altötting 2013. (As a regional compilation for southeast Bavaria, this is one of the essential local books for understanding one's own findings and their classification.)

Kudrna, O. (ed.), *Butterflies of Europe: Aspects of the Conservation of Butterflies in Europe.* Aula-Verlag Wiesbaden, Germany 1986.

Lees, David C. and Alberto Zilli, *Moths: Their Biology, Diversity and Evolution.* Natural History Museum, London 2019.

Lewis-Stempel, John, *Meadowland: The Private Life of an English Field.* Doubleday, London 2014. (A description of the diverse life that exists in English fields, which could also exist or be facilitated in our German fields.)

Louv, Richard, *Das letzte Kind im Wald. Geben wir unseren Kindern die Natur zurück!* Verlag Herder GmbH, Freiburg 2013. (Passionate appeal for the abolition of the utterly useless restrictions on access to nature, so that children can once again become inspired about nature instead of being excluded from it. Unfortunately, our politically influential nature conservation organizations have not reacted.)

McCarthy, Michael, *The Moth Snowstorm: Nature and Joy.* John Murray, London 2015. (One of the loveliest books about experiencing nature, particularly on the subject of moths, that I have ever come across: a fascinating read.)

Muirhead-Thomson, R.C., *Trap Responses of Flying Insects.* Academic Press, London 1991. (Methodical investigation into how insects react to being trapped: very important to avoid drawing the wrong conclusions when evaluating findings!)

Nabokov, Vladimir, *Speak, Memory: An Autobiography Revisited.* Vintage, London 2000. (Autobiography of the author and enthusiastic butterfly collector.)

Oates, Matthew, *In Pursuit of Butterflies. A Fifty-year Affair.* Bloomsbury Natural History, London 2016. (Currently, probably the best description of how intensely people in the United Kingdom observe butterflies and in so doing compile invaluable data: a model for the rest of us in central Europe.)

Owen, J., *The Ecology of a Garden: The First Fifteen Years.* Cambridge University Press, Cambridge 1991. (An astonishing account of the rich variety of wildlife recorded in a suburban garden in the English town of Leicester as a result of a systematic study by the author.)

Parsons, M., 'The changing moth and butterfly fauna of Britain during the twentieth century.' *Entomologist's Rec. J. Var.* 115 (2003): 49–66.

Parsons, M., 'The changing moth and butterfly fauna of Britain – The first decade of the twenty-first century (2000-2009).' *Entomologist's Record and Journal of Variation* 122 (2010): 13–22.

Reichholf, Josef H., 'Untersuchungen zur Biologie des Wasserschmetterlings Nymphula nymphaeata.' *Internationale Revue der Gesamten Hydrobiologie* 55 (1970): 687–728. (Doctoral thesis of the author; I have set out below a selection of other publications that are directly relevant to the subject of this book.)

Reichholf, Josef H., 'Die Massenvermehrung der Gespinstmotte Yponomeuta evonymellus L. (Lepidoptera, Yponomeutidae) im Sommer 1971 am unteren Inn.' *Nachrichtenblatt der Bayerischen Entomologen* 21 (1972): 106–116.

Reichholf, Josef H., 'Die Bedeutung nicht bewirtschafteter Wiesen für unsere Tagfalter.' *Natur und Landschaft* 48 (1973): 80–81.

Reichholf, Josef H., 'Die Feinstruktur der Cuticula hydrophiler und hydrophober Raupen des Wasserschmetterlings *Nymphula nymphaeata* (Lepidoptera: Pyralidae: Nymphulinae).' *Entomol. Germ.* 2 (1976): 258–261.

Reichholf, Josef H., 'Zur Nischenwahl mitterleuropäischer Wasserschmetterling.' *Nachrichtenblatt der Bayerischen Entomologen* 27 (1978): 116–126.

Reichholf, Josef H., *Schmetterlinge beobachten.* BLV Naturführer Munich 1984.

Reichholf, Josef H., *Die Zukunft der Arten.* Beck, Munich 2005.

Reichholf, Josef H. and Miki Sakamoto, 'Die Massenwanderung von Distelfaltern Cynthia cardui Anfang Juni 2003 durch das Bayerische Alpenvorland.' *Atalanta* 36 (2005): 101, 107.

Reichholf, Josef H., *Stadtnatur.* Oekom Verlag, Munich 2007.

Reichholf, Josef H., 'Traubenkirschen-Gespinstmotten Yponomeuta evonymellus in den Auen am unteren Inn: Häufigkeitsentwicklung und Ursache von Massenvermehrungen.' *Mitteilungen der Zoologischen Gesellschaft Braunau* 9 (2008): 273–282.

Reichholf, Josef H., *Der Tanz um das goldene Kalb. Der Ökokolonialismus Europas.* Wagenbach Klaus, Berlin 2011.

Reichholf, Josef H., *Mein Leben für die Natur.* Fischer, Frankfurt 2015.

Reichholf, Josef H., *Das Verschwinden der Schmetterlinge und was dagegen unternommen werden sollte.* Deutsch Wildtier Stiftung, Hamburg 2017.

Reichholf, Josef H., *Schmetterlinge bestimmen in drei Schritten.* Buchverlag, Munich 2017. (Field guide for beginners; a selection of the species that one is most likely to see.)

Reichholf, Josef H., *Schmetterlinge und Vögel im Fokus: Wodurch ändetern sich ihre Häufigkeiten in den letzten Jahrzehnten?* Rundgespräche Forum Ökologie der Bayerischen Akademie der Wissenschaften Bd. 46 (2017): 73–90.

Sage, Walter, *Die Schmetterlinge (Lepidoptera) im Inn-Salzach-Gebiet, Südostbayern. Vorkommen und Veränderungen von 1995 bis 2007. Mitteilungen der Zoologischen Gesellschaft Braunau* 12, Supplement 1–77 (2017). (A record of the 1,121 Lepidoptera species that have been recorded in the Inn-Salzach area of southeast Bavaria to date, with an assessment of changes to occurrence and abundance of species; another study that I drew on for this book.)

Segerer, Andreas H. and Axel Hausmann (eds), *Die Großschmetterlinge Deutschlands*. Heterocera Press, Budapest 2011. (The range of species found in Germany arranged by colour charts.)

Steiner, Axel, Ulrich Ratzel, Morton Top-Jensen and Michael Fibiger, *Die Nachtfalter Deutschlands. Ein Feldführer*. BugBook, Oestermarie, 2014. (The best field guide currently available for macro Lepidoptera, which includes all the species of central Europe.)

Sterling, Phil and Mark Parsons, *Field Guide to the Micro-Moths of Great Britain and Ireland*. British Wildlife Publishing, Gillingham 2012. (The best field guide to the whole of the Microlepidoptera of the British Isles currently available.)

Sterry, P. and Cleave, A., *Collins Complete Guide to British Butterflies and Moths*. Collins, London 2016.

Strong, D.R., J.H. Lawton and Richard Southwood, *Insects on Plants. Community Patterns and Mechanisms*. Harvard University Press, Cambridge, MA 1984. (A key piece of ecological literature; since supplemented and extended through a multitude of scientific publications.)

Tolman, Tom and Richard Lewington, *Die Schmetterlinge Europas und Nordwestafrikas: Alle Tagfalter, über 2000 Arten*. Franckh-Kosmos, Stuttgart 2012. (A field guide, as the name suggests. Original English title: *Collins Butterfly Guide*. London 2008.)

Townsend, Martin and Paul Waring, *Concise Guide to the Moths of Great Britain and Ireland*. Bloomsbury, London 2007.

Tree, Isabella, *Wilding. The Return of Nature to a British Farm*. Picador, London 2018. (A fascinating and inspiring account of how the author and her conservationist husband turned their failing farm in Sussex, southeast England, into an outstanding refuge for wildlife by allowing it to return to a natural condition by ecological management.)

Vane-Wright, Dick, *Butterflies. A Complete Guide to their Biology and Behaviour*. Natural History Museum, London 2015. (Short but extremely comprehensive description of the biology of butterflies; moreover, attractively illustrated.)

Vitousek, Peter M. et al., 'Human alteration of the global nitrogen cycle: causes and consequences.' *Issues in Ecology* I (1997). Ecological Society of America. (This publication established 20 years ago what the agricultural sector is still unwilling to acknowledge: the catastrophic overapplication of fertilizer for which it is responsible.)

Wagner, D.L., 'Moth decline in the Northeastern United States.' *News of the Lepidopterists' Society* 54 (2012): 52–56.

Waring, P. and Townsend, M., *Field Guide to the Moths of Great Britain and Ireland.* British Wildlife Publishing, Rotherwick, Hampshire 2003. (An excellent, beautifully illustrated field guide to the larger moths.)

All translations by Gwen Clayton unless otherwise marked.

Index

Illustrations are indicated by photograph numbers in italics. Species are listed under their common names with the scientific name in parentheses where it is included in the text.